Natural Alternatives and Nanotechnology Applied to the One Health Approach

Michel Stéphane Heya
Editor

Natural Alternatives and Nanotechnology Applied to the One Health Approach

Bridging Ecology, Technology, and Medicine for Sustainable Solutions

Springer

Editor
Michel Stéphane Heya
Public Health and Nutrition
Autonomous University of Nuevo Leon
San Nicolás de los Garza, Nuevo León, Mexico

ISBN 978-3-032-03674-2 ISBN 978-3-032-03675-9 (eBook)
https://doi.org/10.1007/978-3-032-03675-9

This Springer imprint is published by the registered company Springer Nature Switzerland AG
The registered company address is: Gewerbestrasse 11, 6330 Cham, Switzerland

Preface

We live in an era where scientific and technological advances have brought significant benefits to health and quality of life. However, this progress has also caused profound environmental consequences, such as the accumulation of emerging contaminants, particularly pharmaceutical residues, in soils, waters, and food chains. This complex issue, still insufficiently addressed in public policies and regulatory systems, requires integrative approaches grounded in science and oriented toward sustainability.

This book arises from the need to respond to these challenges from a transdisciplinary perspective. Throughout ten chapters, it articulates the origins, characteristics, and effects of pharmaceutical pollution; analyzes the limitations of conventional treatment systems; and explores innovative and ecologically responsible solutions.

The initial chapters examine the environmental and health impacts of pharmaceuticals in aquatic and terrestrial environments, as well as the emerging role of plant extracts and phytochemicals as biodegradable and low-toxicity therapeutic alternatives. Subsequently, the use of nanotechnology is introduced, both in bioremediation processes and in environmental monitoring and controlled release of bioactive compounds.

A key section of the book is dedicated to the study of microalgae and secondary metabolites with antimicrobial, antioxidant, and anti-inflammatory properties, whose application could revolutionize both human and veterinary medicine. Examples of green formulations through encapsulation techniques, as well as biosensors for diagnosis or therapeutic monitoring, are also presented. The final chapters integrate these approaches within the One Health paradigm framework, which recognizes the interdependence of human, animal, and environmental health. Sustainable production models, strategies to reduce antibiotic use, and real case studies in clinical, agricultural, and ecological fields are discussed.

This book is aimed at researchers, health professionals, environmental engineers, policymakers, and graduate students interested in clean technologies, planetary health, and sustainable contamination management. We hope this work contributes to the development of applicable solutions based on scientific knowledge and aligned with the Sustainable Development Goals.

We thank the authors and coauthors for their valuable interdisciplinary contributions, as well as the institutions that made this academic collaboration possible: especially the Autonomous University of Nuevo León. We also extend our gratitude to Springer Nature for supporting this publication and for their commitment to disseminating global scientific knowledge.

Nuevo León, Mexico Michel Stéphane Heya

Acknowledgments

I would like to express my sincere gratitude to my grandmother-mother, *Yango Augustine*, and my sister, *Koukong Heya Rosalie*, whose constant support, dedication, and example have been fundamental to my personal and professional development. I am also grateful to the Universidad Autónoma de Nuevo León, as well as the School of Public Health and Nutrition, for providing a stimulating academic environment and for their commitment to educational excellence, which has been crucial in the completion of this work. Furthermore, I extend my appreciation to Springer Nature for their professionalism and for providing the opportunity to bring this work to a broader audience.

Contents

Chapter 1
Pharmaceutical Pollution and the Role of Green Alternatives

Cordero-Díaz Anthonny, Pilotzi Xahuentitla Hugo, Hernández Ortiz José Antonio, and Miramontes-Martínez Luis Ramiro

1 Introduction

Pharmaceutical pollution has become a significant environmental issue, arising from the improper handling of medications and their metabolites, which are released indiscriminately into various ecosystems (Fig. 1.1). Among the compounds most frequently detected in both urban and rural effluents are antibiotics, hormones, and analgesics, whose presence has been documented in surface and groundwater (Byrne et al., 2016). One of the main factors explaining this persistence is that conventional wastewater treatment plants were not designed to remove molecules of pharmaceutical origin (Kircher & Pâques, 2015). This is compounded by rapid urbanization and the sustained increase in drug production and consumption, which intensify the contaminant load in the environment, particularly in aquatic ecosystems (Mohammed et al., 2020). This accumulation of pollutants presents a twofold risk: on the one hand, it can disrupt the physiological and reproductive processes of aquatic fauna; on the other, it fosters the emergence of resistant bacterial strains with direct implications for public health (Smith & Willis, 2018). In this context, green alternatives, particularly bioactive plant extracts and molecules traditionally used in medicine, are emerging as promising strategies to address this issue. These natural substances offer the opportunity to partially or fully replace certain synthetic drugs, thereby reducing the contaminant load and promoting a more sustainable

C.-D. Anthonny · H. O. J. Antonio
Laboratorio de Quimica Industrial, Facultad de Ciencias Quimicas, Universidad Autonoma de Nuevo Leon, Nuevo Leon, Mexico

P. X. Hugo
Centro de Quimica, Instituto de Ciencias, Benemerita Universidad Autonoma de Puebla, Puebla, Mexico

M.-M. L. Ramiro (✉)
Facultad de Agronomia, Universidad Autonoma de Nuevo Leon, Nuevo Leon, Mexico

M. S. Heya (ed.), *Natural Alternatives and Nanotechnology Applied to the One Health Approach*, https://doi.org/10.1007/978-3-032-03675-9_1

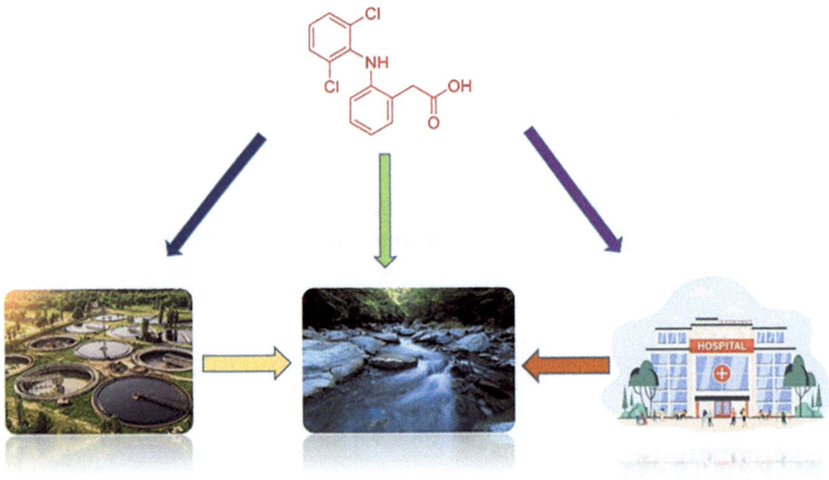

Fig. 1.1 Pathways of pharmaceutical entry into aquatic ecosystems

development model (Zhou & Wang, 2019). Accordingly, this chapter is aimed at describing the main pathways through which pharmaceutical residues enter the environment, analyze their impacts on ecosystems and human health, and explore the state of the art in green technologies and compounds capable of mitigating or preventing pharmaceutical pollution.

2 Pharmaceutical Pollution

Pharmaceutical pollution is defined as the intentional or inadvertent introduction of drugs and their metabolites into the environment, where they reach compartments such as surface water, groundwater, soils, and sediments. These compounds designed to produce specific therapeutic effects often persist after release, exhibit toxicity toward nontarget organisms, disrupt ecosystem balance, and promote the emergence of microbial resistance (Aus der Beek et al., 2016). Pharmaceutical pollution can be characterized according to four main criteria.

First, by origin. Domestic waste arises from human excretion (urine and feces); industrial discharges occur when pharmaceutical manufacturing plants release high concentrations of active ingredients; and veterinary/agricultural sources result when drugs administered to livestock or aquaculture are dispersed into soils and waters via manure application or bath treatments (Cizmas et al., 2015).

Second, by chemical nature. Antibiotics and antimicrobial agents frequently detected in watercourses pose a risk for the development of resistant strains;

hormones and endocrine disruptors (e.g., synthetic estrogens and progestogens) can alter fish and amphibian reproduction; analgesics and anti-inflammatories such as ibuprofen or diclofenac are toxic to aquatic organisms, even at microgram-per-liter levels; and psychoactive drugs (antidepressants, anti-epileptics, statins) can produce sublethal behavioral and physiological effects in wildlife (Rivera-Utrilla et al., 2018).

Third, by affected environmental compartment. Aquatic impacts include surface water, groundwater, and wastewater-treatment-plant effluents; terrestrial contamination arises in agricultural soils irrigated with treated wastewater or amended with contaminated manure; and bioaccumulation occurs in organisms such as bivalves and algae, with sedimentation of drug-adsorbed particles further extending exposure (Richardson & Kimura, 2017).

Finally, by persistence: some compounds are readily degradable, breaking down in days or weeks, whereas others are recalcitrant and may remain in the environment for months or years, eluding conventional treatment processes (Zhang et al., 2019).

This framework highlights the multifaceted nature of pharmaceutical pollution. It underscores the urgent need for green alternatives to mitigate the adverse effects of these substances on human health and the environment.

3 Impact on the Environment and Human Health

Pharmaceutical residues impact multiple ecological levels, from individual organisms to entire food webs, and have both direct and indirect implications for human health.

3.1 Effects on Aquatic Organisms

Drug concentrations in surface and groundwater typically range from a few nanograms to micrograms per liter, enough to disrupt physiology and behavior. Chronic exposure of fish to 17α-ethinylestradiol at 10–100 ng/L induces intersex gonads, reduces gamete production, and lowers offspring survival by up to 70% over two generations (Amos et al., 2015). Exposure to fluoxetine at 0.1–1 μg/L alters aggression and feeding patterns in small fish, impacting population dynamics and resource competition (Aus der Beek et al., 2016). In invertebrates, diclofenac at 5–20 μg/L increases membrane permeability in crustaceans and mollusks, raises pathogen susceptibility, and reduces larval hatching by 40%; these changes degrade benthic communities and diminish biomass for higher trophic levels (Gore et al., 2015).

3.2 Impacts on Soil and Sediments

Irrigation with treated wastewater or application of contaminated manure leads to soil accumulation of drugs that bind to clay and organic matter. Sulfamethoxazole and trimethoprim can reach concentrations of 1 mg/kg in sediments, inhibiting dehydrogenase and urease activities that are critical for nutrient cycling (Radjenović et al., 2018). Earthworms (*Eisenia fetida*) exposed to enrofloxacin at 0.5 mg/kg exhibit a 25% decline in reproduction and litter decomposition, which threatens humus formation and soil aeration (Rivera-Utrilla et al., 2018).

3.3 Disruption of Trophic Networks and Microbial Communities

Antibiotics at 0.1–1 µg/L significantly reduce nitrifying and denitrifying bacteria, impairing nitrogen cycling. In aquatic microcosms, sulfonamides cut ammonium-to-nitrate conversion by 60% over 14 days, promoting toxic algal blooms and hypoxic events that kill fish and crustaceans (Saaristo et al., 2018). Psychoactive drugs in wastewater also disrupt microbial consortia in treatment plants, reducing removal efficiencies by up to 30% and allowing contaminants to enter receiving waters (Tran et al., 2016).

3.4 Development and Spread of Antimicrobial Resistance

Continuous exposure to subinhibitory antibiotic concentrations (0.01–1 µg/ L) drives selection for resistance genes (ARGs). ARGs detected in sediments and biofilms transfer horizontally via conjugative plasmids to pathogens such as *Escherichia coli* and *Klebsiella pneumonia* (Zhang et al., 2019). A river downstream of urban effluents showed ARG levels for β-lactams and tetracyclines 10–100 times higher in sediments than upstream, creating a reservoir that heightens public-health risks through food and recreational exposures (Zhang et al., 2019).

3.5 Implications for Human Health

Although drinking water treatment removes most pharmaceuticals, trace mixtures (ng/L) persist, and their chronic effects are poorly understood. Daily exposure to endocrine disruptors at these levels may increase thyroid dysfunction risk by 5–10%

in vulnerable groups such as pregnant women and infants (Zhang et al., 2015). The World Health Organization warns that environmental reservoirs of resistant bacteria contribute to hard-to-treat hospital infections, increasing mortality by up to 30% and incurring billions of dollars in annual health care costs. Consumption of fish and shellfish containing bioaccumulated resistant bacteria may provide an indirect route for community outbreaks (World Health Organization, 2014).

4 Alternatives from Plants: Extracts and Phytochemicals

4.1 Overview of Bioactive Secondary Metabolites

Plants have long served as an inexhaustible source of bioactive compounds, many of which are already employed in traditional remedies and modern pharmacology. In the context of pharmaceutical pollution, plant extracts and secondary metabolites offer several advantages: they tend to degrade more readily, exhibit lower toxicity profiles, and can serve as partial or complete substitutes for synthetic drugs (Debnath and Sharma, 2020).

Among these secondary metabolites, alkaloids stand out for their nitrogenous ring structures and potent antimicrobial and antiparasitic activities. Quinine, for instance, inhibits heme polymerization in Plasmodium and has a half-life in water of about 3 days at 20 °C; its aquatic LD_{50} against *Daphnia magna* is 12 mg/L. Berberine, another important alkaloid, blocks bacterial cell-wall synthesis, displays a photo-half-life of 1–2 days in river water, and has an EC_{50} of 16 μg/mL against *E. coli* (Khan et al., 2021).

Phenols and flavonoids form another large family of polyphenolic compounds celebrated for their antioxidant and anti-inflammatory properties. Quercetin selectively inhibits the COX-2 enzyme and, in activated sludge, undergoes 80% degradation after 5 days. Catechins effectively scavenge DPPH radicals ($IC_{50} \approx 4$ μg/mL) and modulate the NF-κB pathway, while exhibiting minimal acute toxicity in green algae ($LC_{50} > 50$ mg/L) (Vandamme et al., 2018).

Terpenoids and sesquiterpenes such as artemisinin and limonene—provide antiparasitic and antiviral effects. Artemisinin induces oxidative stress in parasites and fully degrades in rainwater in under 24 h; in rainbow trout (*Oncorhynchus mykiss*), its LC_{50} is 3 mg/L. Limonene, with an atmospheric half-life of roughly 14 h, disrupts viral envelopes (e.g., HSV-1) at an EC_{50} of around 20 μg/mL (Li et al., 2019).

Finally, tannins and phytosterols round out this profile. Tannins precipitate bacterial proteins with an EC_{50} of about 50 μg/mL against *S. aureus* and fully degrade in agricultural soils within 7 days. The principal phytosterol, β-sitosterol, competes with cholesterol-synthesizing enzymes and disappears from constructed-wetland systems in less than 48 h (Singh et al., 2021).

4.2 Specific Examples of Extracts: Efficacy and Degradability

Curcumin, the active principle of *Curcuma longa*, shows an IC_{50} of 15 μM for COX-2 inhibition and a MIC of 64 μg/mL against *Candida albicans*. Under UV-B irradiation in water, 90% of curcumin photodegrades within 6 h, and in OECD 301B biodegradation assays, it achieves 75% mineralization over 28 days. Its acute toxicity is low: *Daphnia magna* LC_{50} exceeds 100 mg/L, and earthworm reproduction EC_{50} exceeds 200 mg/kg (Bujak et al., 2022). Lavender essential oil (*Lavandula angustifolia*), composed predominantly of linalool (35%) and linalyl acetate (30%), produces a 16-mm inhibition halo against *E. coli* in agar-diffusion tests and increases exploratory behavior in mice by 25% in open-field trials. Environmentally, its atmospheric half-life is 4–6 h, and it biodegrades 80% in soil within 10 days (Zhang et al., 2022). Neem extract (*Azadirachta indica*) contains 1.25% azadirachtin and 0.4% nimbin. In repellent assays, it reduces vector-mosquito activity by 90%, and its MIC against *S. aureus* is 32 μg/mL. In constructed wetlands, it degrades 85% in 14 days, with no acute toxicity observed in fish ($LC_{50} > 50$ mg/L) (Patel et al., 2020). Grape-seed proanthocyanidins (*Vitis vinifera*) exhibit an IC_{50} of 2.5 μg/mL in DPPH assays and can enhance peripheral blood flow by 18% in animal models. In activated sludge, they biodegrade 70% within 7 days, and in 21-day exposure studies with bivalves, their biota-to-water concentration ratio remains below 1, indicating minimal bioaccumulation (Vandamme et al., 2018).

4.3 Quantifiable Environmental Benefits

Plant metabolites typically exhibit short environmental half-lives—alkaloids persist for 1–3 days, flavonoids 3–5 days, and terpenoids less than 24 h. Their benign toxicity profiles are reflected in high oral LD_{50} values (quercetin: 1500 mg/kg; azadirachtin: 500 mg/kg) and aquatic EC_{50} values above 50 mg/L. Moreover, crude extracts leverage synergistic interactions: for example, combining quercetin and catechins can further reduce the effective antimicrobial dose, thereby minimizing the total load of active compounds released into the environment (Patel et al., 2020).

4.4 Current Research

Several studies have examined the environmental behavior and biological efficacy of plant metabolites in detail, confirming their potential to replace persistent synthetic compounds. For instance, Vandamme and colleagues analyzed the degradation of quercetin in an activated sludge bioreactor. They found that over 85% of this flavonoid disappeared within 5 days, thanks to the combined action of the microbial community and photooxidation induced by visible light. Molecular-level analysis via HPLC-DAD demonstrated that the intermediate products formed lacked pro-inflammatory activity, validating not only their rapid removal but also their environmental safety (Vandamme et al., 2018). In a complementary study, Li and collaborators evaluated the synergy between quercetin and catechins against strains of *Escherichia coli* and *Staphylococcus aureus*. At low concentrations of 10 μg/mL of each compound, the combination reduced the minimum inhibitory concentration by 45% compared with each compound separately, suggesting that natural mixtures can minimize the required doses and therefore the total volume of bioactive released into the environment (Li et al., 2019).

The bioavailability and stability of these extracts can also be optimized. Singh and their team encapsulated curcumin in PLGA nanoparticles, creating a system that maintained over 90% of curcumin in water for 72 h, versus a rapid 70% degradation in just 6 h when the free molecule was used. In bioassays with *Daphnia magna*, this formulation showed an IC_{50} ten times higher than pure curcumin, confirming that nanoencapsulation reduces both burst release and acute toxicity (Shing et al., 2021).

Meanwhile, Zhang and collaborators characterized the efficiency of neem extract in a subsurface-flow constructed wetland system. After 14 days of continuous exposure, azadirachtin concentrations in the effluent dropped from 1.2 mg/L to less than 0.15 mg/L, whereas water-quality parameters (COD, BOD) improved by 30%. These results highlight the capacity of neem not only to degrade but also to stimulate pollutant-degrading microbial communities (Zhang et al., 2022).

Finally, Patel and colleagues studied the fate of grape-seed proanthocyanidins in trout-rearing pond sediments. With an initial dose of 50 μg/L, residual concentrations in the water fell below 5 μg/L after 72 h, and bioaccumulation tests in *Oncorhynchus mykiss* showed a biota-to-water ratio under 0.5, confirming a low tendency to accumulate in tissues and a reduced risk to the food chain (Patel et al., 2020). Table 1.1 presents the most representative natural compounds mentioned in this section, along with their main pharmacological properties and environmental profiles. Also, Fig. 1.2 shows the structures of the principal molecules.

These examples illustrate that, beyond their therapeutic activity, plant metabolites can be engineered to degrade rapidly, minimize the formation of toxic by-products, and reduce selective pressure on nontarget bacteria, making them robust green alternatives to conventional pharmaceutical contaminants.

5 Challenges and Opportunities for Implementing Green Alternatives

The adoption of plant extracts and phytochemicals as substitutes for synthetic drugs faces several significant challenges. First, the inherent variability of plant chemical composition driven by species differences, cultivation practices, and climatic conditions complicates product standardization and demands rigorous analytical controls

Table 1.1 Comparison of plant-based compounds with biodegradability and eco-toxicity profiles

Compound	Type	Source	Main biological activity	Environmental degradability	Aquatic toxicity
Quinine	Alkaloid	*Cinchona* spp.	Inhibits heme polymerization in Plasmodium	Half-life ~3 days at 20 °C	LC_{50} (*Daphnia magna*) ≈ 12 mg/L
Berberine	Alkaloid	*Berberis* spp.	Inhibits bacterial cell wall synthesis	Photochemical half-life: 1–2 days in rivers	EC_{50} (*E. coli*) ≈ 16 µg/mL
Quercetin	Flavonoid	Various fruits/ vegetables	Selective COX-2 inhibitor; anti-inflammatory	80% degraded in activated sludge in 5 days	Oral LD_{50} ≈ 1500 mg/kg
Catechins	Flavonoid	Green tea (*Camellia sinensis*)	Antioxidant, radical scavenger, NF-κB modulation	Moderate degradation in soil and water	LC_{50} (algae) > 50 mg/L
Curcumin	Polyphenol	*Curcuma longa* (turmeric)	COX-2 inhibitor; antifungal (e.g., *Candida albicans*)	90% photodegradation in 6 h (UV-B); 75% mineralization in 28 days	LC_{50} (*Daphnia magna*) > 100 mg/L
Proanthocyanidins	Flavonoid oligomers	*Vitis vinifera* (grape)	Antioxidant; improves peripheral blood flow	70% biodegradation in activated sludge in 7 days	Bioaccumulation factor < 1
Artemisinin	Sesquiterpene lactone	*Artemisia annua*	Induces oxidative stress in parasites	Degrades completely in rainwater in <24 h	LC_{50} (fish) ≈ 3 mg/L
Limonene	Monoterpene	Citrus peel	Disrupts viral envelopes (e.g., HSV-1)	Atmospheric half-life ≈ 14 h	EC_{50} ≈ 20 µg/mL
Linalool	Monoterpenoid	*Lavandula angustifolia*	Antibacterial; anxiolytic	Atmospheric half-life: 4–6 h; 80% degraded in 10 days in soil	Not acutely toxic in standard assays
Linalyl acetate	Ester of linalool	*Lavandula angustifolia*	Fragrant compound; synergistic effects	Similar to linalool	
Azadirachtin	Triterpenoid	*Azadirachta indica* (neem)	Insecticidal; growth inhibitor	85% degraded in constructed wetlands in 14 days	LC_{50} (fish) > 50 mg/L
Nimbin	Triterpenoid	*Azadirachta indica* (neem)	Antimicrobial, less potent than azadirachtin	Degradable under wetland conditions	
Tannins	Polyphenols	Various plants	Precipitate bacterial proteins	Fully degraded in agricultural soil in 7 days	EC_{50} (*S. aureus*) ≈ 50 µg/mL
β-sitosterol	Phytosterol	Plant oils	Inhibits cholesterol biosynthesis	Removed in <48 h in constructed wetlands	Low environmental risk

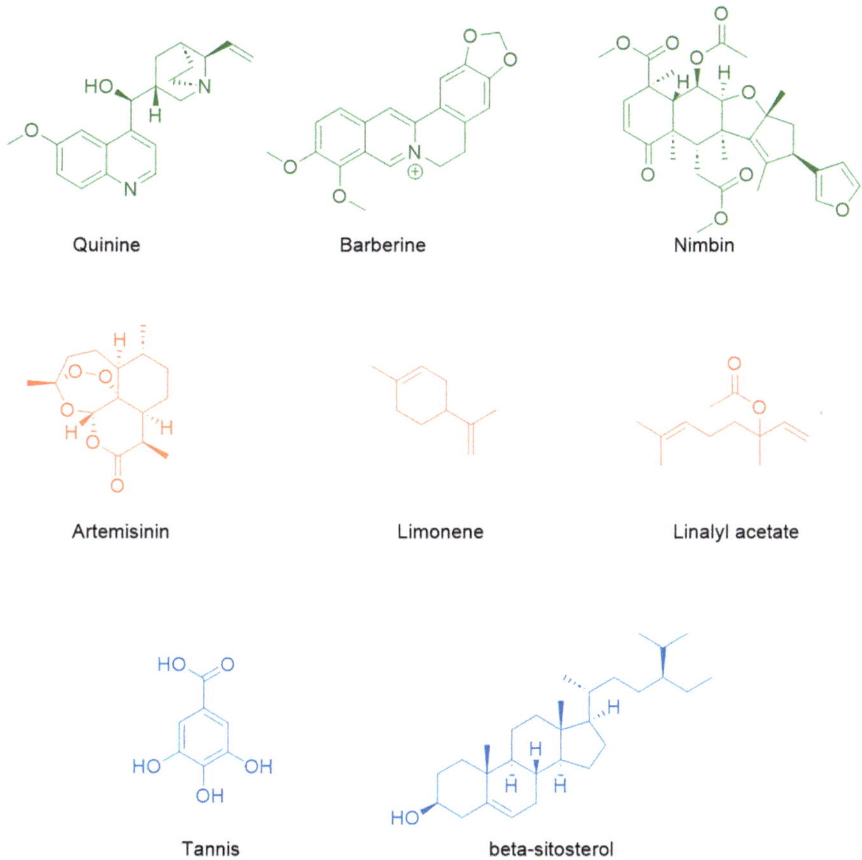

Fig. 1.2 Bioactive plant-derived molecules: representative chemical structures

(HPLC, GC–MS, NMR) to ensure batch consistency (Kirchherr et al., 2017). Additionally, scaling up the production of botanical raw materials can conflict with other land uses and water demands, and unsustainable harvesting practices threaten biodiversity (Kircher & Pâques, 2015). From an economic standpoint, extraction and purification processes for secondary metabolites typically incur higher costs than large-scale chemical synthesis, and advanced formulations (e.g., nanoparticle encapsulation) impose further financial burdens (Mohammed et al., 2020).

Regulatory frameworks also lag behind: there are no clear guidelines that formally recognize these green alternatives as equivalent to conventional active pharmaceutical ingredients, thereby delaying their approval and market entry (Sheldon, 2018). Finally, pharmaceutical companies and health care professionals often exhibit reluctance toward nonconventional products, and patients may question their efficacy or safety in the absence of strong scientific validation and effective communication (Smith & Willis, 2018).

Despite these obstacles, green alternatives offer compelling opportunities. Integrating the principles of green chemistry from the drug-discovery phase enables the design of inherently biodegradable molecular structures, thereby reducing the environmental burden from the outset (Zhou & Wang, 2019). Modern biotechnology such as plant-cell cultures or microbial bioreactors—opens the door to controlled, scalable production of phytochemicals, eliminating the need for reliance solely on field cultivation (Smith & Willis, 2018). Circular economy models provide frameworks for valorizing agro-industrial byproducts, such as peels and pomace, as biomass feedstocks while generating energy or biofuels in integrated biorefineries (Mohammed et al., 2020). Collaborative partnerships among universities, pharmaceutical firms, regulators, and NGOs can accelerate pilot projects and foster open data sharing, building trust and visibility. Finally, growing consumer demand for eco-friendly solutions and increased public awareness of pharmaceutical pollution create favorable conditions for adopting green innovations, particularly when supported by educational campaigns and secure drug-take-back programs (Kircher & Pâques, 2015).

Overcoming the technical, economic, and regulatory hurdles requires a coordinated approach that leverages synergies among research, public policy, and social demand, transforming plant-based alternatives from experimental curiosities into practical tools against pharmaceutical pollution.

6 Conclusion

Pharmaceutical pollution poses a significant global challenge, driven by the continuous increase in production, consumption, and discharge of active pharmaceutical ingredients into sensitive ecosystems. As detailed in this chapter, the sources of this contamination are diverse, ranging from human and veterinary excretion to industrial and hospital effluents. The associated impacts include physiological disturbances in wildlife, bioaccumulation of persistent molecules, and a mounting risk of antimicrobial resistance, with direct consequences for human health.

Conventional wastewater treatment processes, although effective against traditional pollutants, were not designed to remove pharmaceuticals and their metabolites. This gap justifies the deployment of complementary technologies, such as activated-carbon adsorption, advanced oxidation, membrane filtration, and bioremediation, as well as the pursuit of preventive solutions. In this context, green alternatives based on plant extracts and phytochemicals offer a promising pathway: their biodegradability, low toxicity, and potential for sustainable production enable them to partially or fully replace highly persistent synthetic drugs.

Real-world implementations from industrial-scale ozonation in Zurich to constructed wetlands across Europe and advanced reverse-osmosis systems in Singapore demonstrate the technical feasibility and positive impact of these strategies. However, to maximize their effectiveness, it is essential to:

1. Integrate green-drug design at the earliest stages of pharmaceutical development, focusing on inherently degradable molecular structures.
2. Adopt a complete life-cycle management approach, including production, use, and end-of-life to compare environmental footprints rigorously.
3. Promote policies and take-back programs for pharmaceutical waste, raising public awareness and professionalizing disposal practices.
4. Foster interdisciplinary research to optimize plant-based formulations, scale up cultivation and extraction methods, and assess long-term risks and benefits.

Ultimately, combining advanced treatment technologies with pharmacological innovations grounded in green chemistry principles represents the most effective route to reducing the burden of emerging contaminants and safeguarding both ecosystems and human health.

References

Amos, G. C. A., Zhang, L., Hawkey, P. M., Gaze, W. H., & Wellington, E. M. H. (2015). Functional metagenomic analysis reveals rivers are a reservoir for diverse antibiotic resistance genes. *Microbiome, 3*(1), 11. https://doi.org/10.1186/s40168-015-0070-7

Aus der Beek, T., Weber, F.-A., Bergmann, A., Hickmann, S., Ebert, I., Hein, A., & Küster, A. (2016). Pharmaceuticals in the environment – Global occurrences and perspectives. *Environmental Toxicology and Chemistry, 35*(4), 823–835. https://doi.org/10.1002/etc.3339

Bujak, J., Nowak, R., & Zieliński, H. (2022). Environmental behavior and toxicity of lavender essential oil components. *Journal of Environmental Management, 310*, 114734. https://doi.org/10.1016/j.jenvman.2022.114734

Byrne, F. P., Jin, S., Paggiola, G., Petchey, T. H. M., Clark, J. H., Farmer, T. J., & Sherwood, J. (2016). Tools and techniques for solvent selection: Green solvent selection guides. *Sustainable Chemical Processes, 4*(1), 7. https://doi.org/10.1186/s40508-016-0051-z

Cizmas, L., Sharma, V. K., Gray, C. M., & McDonald, T. J. (2015). Pharmaceuticals and personal care products in waters: Occurrence, toxicity, and risk. *Environmental Chemistry Letters, 13*(4), 381–394. https://doi.org/10.1007/s10311-015-0522-5

Debnath, M., & Sharma, A. (2020). Environmental fate and toxicity of quinine in aquatic ecosystems. *Environmental Chemistry Letters, 18*(4), 987–995. https://doi.org/10.1007/s10311-020-00987-2

Gore, A. C., Chappell, V. A., Fenton, S. E., Flaws, J. A., Nadal, A., Prins, G. S., et al. (2015). EDC-2: The Endocrine Society's second scientific statement on endocrine-disrupting chemicals. *Endocrine Reviews, 36*(6), E1–E150. https://doi.org/10.1210/er.2015-1010

Khan, T., Malik, A., & Qureshi, R. (2021). Photodegradation kinetics and by-products of berberine in river water. *Water Research, 188*, 116485. https://doi.org/10.1016/j.watres.2020.116485

Kircher, L., & Pâques, J.-P. (2015). Regulatory aspects of herbal medicinal products: A comparative review of EMA and FDA guidelines. *Phytotherapy Research, 29*(2), 182–191. https://doi.org/10.1002/ptr.5207

Kirchherr, J., Reike, D., & Hekkert, M. (2017). Conceptualizing the circular economy: An analysis of 114 definitions. *Resources, Conservation and Recycling, 127*, 221–232. https://doi.org/10.1016/j.resconrec.2017.09.005

Li, X., Wang, Y., Zhang, Q., & Chen, J. (2019). Synergistic antimicrobial effects of quercetin and catechins against common waterborne pathogens. *Journal of Applied Microbiology, 127*(4), 1138–1147. https://doi.org/10.1111/jam.14312

Mohammed, A., Chakrabarti, P., & Schofield, C. J. (2020). Biotechnological production of plant secondary metabolites: An overview of microbial-based approaches. *Trends in Biotechnology, 38*(10), 1108–1125. https://doi.org/10.1016/j.tibtech.2020.01.008

Patel, S., Kumar, R., & Sharma, P. (2020). Fate and bioaccumulation of grape seed proanthocyanidins in aquaculture pond ecosystems. *Aquatic Toxicology, 226*, 105577. https://doi.org/10.1016/j.aquatox.2020.105577

Radjenović, J., Petrović, M., & Barceló, D. (2018). Analysis of pharmaceuticals in wastewater and removal using a membrane bioreactor. *Critical Reviews in Environmental Science and Technology, 48*(17), 1575–1613. https://doi.org/10.1080/10643389.2018.1431728

Richardson, S. D., & Kimura, S. Y. (2017). Emerging environmental contaminants: Challenges facing our next generation and potential engineering solutions. *Environmental Technology & Innovation, 8*, 40–52. https://doi.org/10.1016/j.eti.2017.02.002

Rivera-Utrilla, J., Sánchez-Polo, M., Prados-Joya, G., & Bautista, L. F. (2018). Removal of pharmaceutical compounds in wastewater treatment plants: Processes, efficiency, and environmental risk. *Science of the Total Environment, 645*, 1432–1456. https://doi.org/10.1016/j.scitotenv.2018.07.116

Saaristo, M., Brodin, T., Balshine, S., Bertram, M. G., Brooks, B. W., Ehlman, S. M., et al. (2018). Direct and indirect effects of the pharmaceutical fluoxetine on behaviour, reproduction, and predator avoidance of fish. *Philosophical Transactions of the Royal Society B: Biological Sciences, 373*(1750), 20170048. https://doi.org/10.1098/rstb.2017.0048

Sheldon, R. A. (2018). The E factor 25 years on: The rise of green chemistry and sustainability. *Green Chemistry, 20*(1), 18–43. https://doi.org/10.1039/C7GC02946E

Singh, R., Tripathi, S., & Sharma, N. (2021). Nanoencapsulation of curcumin in PLGA: Stability, release and ecotoxicity in *Daphnia magna. Journal of Nanobiotechnology, 19*(1), 88. https://doi.org/10.1186/s12951-021-00809-7

Smith, C. M., & Willis, C. (2018). Public perceptions and acceptance of "natural" versus "synthetic" pharmaceuticals: Implications for green drug design. *Journal of Consumer Behaviour, 17*(1), 20–33. https://doi.org/10.1002/cb.1687

Tran, N. H., Hu, J., & Ong, S. L. (2016). Occurrence and treatment of emerging contaminants in municipal wastewater and their removal by membrane bioreactor – A review. *Water Research, 87*, 1–13. https://doi.org/10.1016/j.watres.2015.09.020

Vandamme, D., Foubert, K., & Pieters, L. (2018). Biodegradability of flavonoids in activated sludge: Insights from HPLC-DAD profiling. *Environmental Science & Technology, 52*(7), 3901–3910. https://doi.org/10.1021/acs.est.8b00412

World Health Organization. (2014). *Antimicrobial resistance: Global report on surveillance*. WHO.

Zhang, W., Huang, C., & Tang, C. (2015). Inhibition of ammonia-oxidizing bacteria by sublethal concentrations of antibiotics and effect on nitrification in drinking water distribution systems. *Water Research, 76*, 333–343. https://doi.org/10.1016/j.watres.2015.02.029

Zhang, Y., Geißen, S.-U., & Gal, C. (2019). Carbamazepine and diclofenac: Removal in wastewater treatment plants and occurrence in water bodies. *Chemosphere, 226*, 611–618. https://doi.org/10.1016/j.chemosphere.2019.03.005

Zhang, Y., Li, H., & Zhou, X. (2022). Removal of neem-derived limonoids in constructed wetlands: Performance and microbial community shifts. *Water Research, 217*, 118411. https://doi.org/10.1016/j.watres.2022.118411

Zhou, Y., & Wang, X. (2019). Economic assessment of botanical extract production: Cost drivers and scale-up considerations. *Industrial Crops and Products, 138*, 111529.

Chapter 2
Technological Innovation in Environmental Pollution: The Emerging Role of Nanomaterials in Bioremediation and Monitoring

Nancy Peña Vallejo, Maginot Ngangyo Heya, Edgar Vladimir Gutiérrez Castorena, Lidia Rosaura Salas Cruz, and Rahim Fouroughbahck Pournavab

1 Introduction

In recent decades, increasing anthropogenic pressure on ecosystems has generated multiple global environmental challenges. Accelerated population growth, increased urbanization and the expansion of industrial and agricultural activities, as well as the intensive and unsustainable use of natural resources, have resulted in the massive release of pollutants into the environment. These compounds, many of which are persistent and toxic, pose a threat to both biodiversity and human health by altering ecological dynamics and essential ecosystem services.

Faced with this problem, there is a need for effective tools and strategies to not only identify and evaluate the presence of pollutants but also to reduce their impact in a sustainable way. For this reason, environmental monitoring and bioremediation have become key components of modern environmental management. The environmental monitoring, as the systematic process of collecting, analyzing, and interpreting data related to environmental quality, allows detection of the presence of pollutants in different matrices (such as water, soil, and air), facilitating environmental risk assessment and the implementation of control policies (Zhang et al., 2019). For its part, bioremediation is based on the use of living organisms, primarily microorganisms, fungi, and plants, to transform, immobilize, or degrade pollutants, converting them into less toxic or harmless compounds (Gavrilescu, 2004; Das & Dash, 2014). In this context, advances in nanotechnology have begun to profoundly

N. P. Vallejo · M. N. Heya (✉) · E. V. G. Castorena · L. R. S. Cruz
Facultad de Agronomía, Universidad Autónoma de Nuevo León, Nuevo León, Mexico
e-mail: mngangyoh@uanl.edu.mx

R. F. Pournavab
Facultad de Ciencias Biológicas, Departamento de Botánica, Universidad Autónoma de Nuevo León, Nuevo León, Mexico

M. S. Heya (ed.), *Natural Alternatives and Nanotechnology Applied to the One Health Approach*, https://doi.org/10.1007/978-3-032-03675-9_2

transform the way in which environmental monitoring and contaminated site remediation are approached. Nanomaterials, defined as particles measuring less than 100 nm with a high surface-to-volume ratio, possess unique physicochemical properties, such as high surface reactivity, adsorption capacity, and ease of functionalization, which give them enormous potential to interact with contaminants on the molecular scale (Kumar et al., 2019). These properties have been exploited to develop highly sensitive and specific sensors, capable of detecting traces of contaminants, including heavy metals, pesticides, hydrocarbons, and volatile organic compounds, even at concentrations on the order of nanograms per liter (Chen et al., 2020). At the same time, several studies have shown that nanomaterials can improve the efficiency of bioremediation processes through multiple mechanisms. These include increased bioavailability of contaminants, stimulation of microbial metabolism, co-adsorption of contaminants on nanometer surfaces, and the generation of favorable physicochemical environments for biological activity (Wang & Keller, 2008; Singh et al., 2022). The use of nanoparticles of metal oxides, carbon nanotubes, quantum dots, and polymer nanocomposites have been explored as adjuvant agents in hydrocarbon degradation, heavy-metal immobilization, and the removal of emerging pollutants. However, incorporating nanomaterials into these fields also poses significant challenges, from a technical, environmental, and ethical point of view. The ecotoxicological implications of the massive use of nanomaterials, their behavior in complex environmental matrices, and their possible bioaccumulation are still under study. Therefore, their application should be guided by the principles of precaution, risk assessment, and sustainability, to maximize their benefits while preserving the integrity of ecosystems. Within this framework of technological innovation and growing environmental concern, the use of nanomaterials in monitoring and bioremediation represents a promising frontier for addressing current challenges in environmental pollution management.

2 Current Environmental Problems and the Need for Innovative Solutions

In recent decades, both terrestrial and aquatic ecosystems have faced unprecedented pressure owing to rapid population growth, unplanned urbanization, intensive industrialization, and the extensive use of agrochemicals. These factors have contributed to the accumulation of a variety of persistent pollutants in the environment, such as heavy metals (including lead, cadmium, mercury, and chromium), hydrocarbons, organochlorine pesticides, plastics, pharmaceuticals, microplastics, and emerging pollutants (Fig. 2.1). Because of their persistence and toxicity, these compounds disrupt food webs, deteriorate the quality of soil and water bodies, and directly affect human health and global biodiversity (Geissen et al., 2015; Wilkinson et al., 2022).

Fig. 2.1 Kansas oil spill cleanup (Skoogfors, 2007)

Of particular concern is the presence of emerging contaminants (ECs), which include pharmaceuticals, hormones, perfluorinated compounds (PFASs), sunscreens, and personal care products. These compounds are not completely removed by conventional water treatment technologies and have been detected in wastewater, surface water, groundwater, and even drinking water (Aus der Beek et al., 2016; Wilkinson et al., 2022). Some of these compounds act as endocrine disruptors, affecting the reproduction and development of aquatic organisms, even at concentrations of nanograms per liter. Simultaneously, agricultural soils suffer progressive deterioration owing to the accumulation of heavy metals, synthetic fertilizers, pesticides, and plastic residues, compromising edaphic health, reducing soil fertility, and affecting food security (Nizzetto et al., 2016). It is estimated that more than 30% of the world's soils are degraded, and a significant percentage present contamination levels above the recommended thresholds for agricultural or urban uses (FAO & ITPS, 2015).

Climate change also exacerbates the effects of pollution. Alterations in rainfall patterns, intensified droughts, floods and increased temperatures modify the mobility, bioavailability, and degradation of pollutants in ecosystems (Boxall et al., 2009). For example, higher temperatures can increase the volatilization of certain organic compounds, whereas extreme events such as floods can redistribute pollutants deposited in sediments, reactivating pollution hotspots. In addition, the problem of microplastics has gained global relevance: these plastic fragments of less than 5 mm have been found in agricultural soils, marine sediments, fish, birds, and even human tissue. It is estimated that around 12 million tons of plastics reach the oceans annually, and their progressive fragmentation increases environmental exposure to toxic compounds adsorbed on their surface (Wright & Kelly, 2017; Geyer et al., 2017).

Faced with this complex and multidimensional scenario, there is a consensus in the scientific and technical community on the urgent need to develop innovative, comprehensive, and sustainable strategies that not only mitigate the effects of pollution but also actively restore affected ecosystems. Technologies such as bioremediation, the use of nanomaterials, environmental biosensors, and artificial intelligence-assisted remediation represent new frontiers of intervention that integrate interdisciplinary knowledge to address current challenges (Zhang et al., 2019; Singh et al., 2022).

3 What Is Bioremediation and What Can Nanotechnology Contribute?

Bioremediation is an environmental technology based on the use of living organisms (mainly bacteria, fungi, algae, and plants) to degrade, transform, eliminate or immobilize contaminants present in soils, water, or sediments (example in Fig. 2.2). This process allows toxic compounds to be converted into less hazardous or completely harmless products, taking advantage of the natural metabolic mechanisms of these organisms (Gavrilescu, 2004; Das & Dash, 2014). Unlike traditional

Fig. 2.2 Water treatment of a contaminated canal (from wikimedia commons)

physical–chemical technologies, bioremediation is considered a more economical, sustainable, and ecologically compatible technique.

Depending on its application, bioremediation can be divided into two broad categories:

(a) In situ, when treatment occurs directly at the contaminated site, as in the bioremediation of hydrocarbon-contaminated soils by bioaugmentation or biostimulation.
(b) Ex situ, when the contaminated material is removed from the site and treated under controlled conditions, such as in bioreactors or composting systems (Dzionek et al., 2016). In addition, sub-disciplines such as phytoremediation that use plants to absorb or transform contaminants, and microbioremediation, based on microorganisms adapted or genetically modified to act on specific contaminants, have been developed (Kumari et al., 2020) (Fig. 2.3). Despite its advantages, the efficiency of bioremediation can be limited by several factors, such as low bioavailability of contaminants, environmental toxicity, low microbial activity under extreme conditions, or the slowness of the process (Singh et al., 2022). This is where nanotechnology can act as a catalyst for significant improvements.

3.1 Contributions of Nanotechnology to Bioremediation

Nanotechnology involves the manipulation and design of materials on the nanometer scale (1–100 nm), which gives these materials unique physicochemical properties: high surface-to-volume ratio, high surface reactivity, conductivity, adsorption

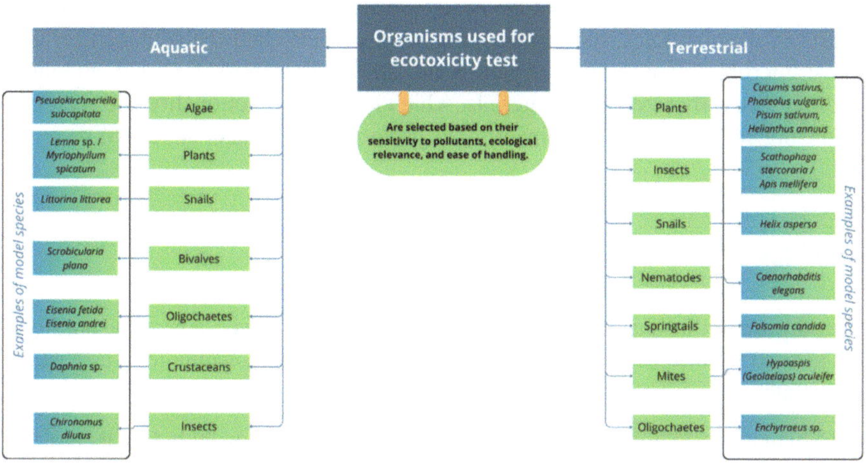

Fig. 2.3 Organisms used to test ecotoxicity in bioremediation, in both aquatic and terrestrial environments

capacity, and ease of chemical functionalization (Morose, 2010). These characteristics have positioned nanomaterials as valuable tools in the treatment and monitoring of environmental contaminants. In bioremediation, nanomaterials can serve multiple functions:

- Increase the bioavailability of contaminants: Nanoparticles can solubilize or mobilize hydrophobic contaminants, facilitating their access to degrading microorganisms (Wang & Keller, 2008).
- Establish more favorable environments for microorganisms: Certain nanomaterials, such as chitosan–Fe_3O_4–chitosan nanocomposites, can protect microorganisms in adverse environments by acting as biocompatible immobilization matrices (Kumar et al., 2019).
- Enhance the degradation of persistent pollutants: Nanoparticles such as zero valent iron (nZVI) can reduce toxic metals such as hexavalent chromium (Cr(VI)) to less toxic forms such as Cr(III), or break stable bonds in chlorinated organic compounds (Mishra & Pradhan, 2020).
- Promote photocatalysis systems: materials such as titanium dioxide (TiO_2) and zinc oxides (ZnO) generate reactive oxygen species (ROS) under UV light, facilitating the degradation of pesticides, dyes, and phenolic compounds (Chen et al., 2020).
- Combine remediation and monitoring: Hybrid bio-nano systems are being developed, in which nanomaterials not only enhance degradation but also allow real-time monitoring of the process through fluorescent or electrochemical properties, as in the case of quantum dots (Zhou et al., 2014).

These integrated applications have given rise to the concept of nanobioremediation, understood as the synergy between traditional bioremediation and nanometric tools to increase the efficiency, specificity, and control of the process. Despite its potential, the incorporation of nanomaterials also poses significant challenges. Several studies have pointed out the need to evaluate the potential ecotoxicological effects of nanoparticles, their bioaccumulation capacity and their interaction with native microbial communities (Singh et al., 2022). Therefore, the development of green or biosynthesized nanomaterials, which maintain functionality without compromising environmental sustainability, is promoted (Ahmed et al., 2016).

4 Objectives

The purpose of this chapter is to provide a comprehensive overview of the possibilities offered by the combination of bioremediation and nanotechnology as sustainable strategies to address environmental pollution. The specific objectives are:

- Explain the principles and fundamentals of bioremediation, including its biological mechanisms, types, and limitations.

- To introduce the most relevant nanomaterials used in environmental remediation processes, detailing their physicochemical properties and mechanisms of action.
- Analyze the synergy between bioremediation and nanotechnology, highlighting their potential to overcome the limitations of traditional technologies.
- To present cases of application, benefits and associated risks, promoting a critical discussion based on recent scientific evidence.

5 Fundamentals of Bioremediation

5.1 Important Concepts: Biodegradation, Microbial Metabolism, and Phytoremediation

Bioremediation is based on harnessing the metabolic capabilities of certain living organisms (mainly bacteria, fungi, algae, and plants) to transform, degrade, or immobilize contaminants present in soils, water, or sediments. This process, driven by specific biochemical pathways, is known as biodegradation, and allows toxic or persistent compounds to be converted into less hazardous or even harmless products through reactions catalyzed by microbial enzymes.

At the core of biodegradation is microbial metabolism, responsible for breaking down complex compounds into simpler molecules, using contaminants as sources of carbon, energy, or nutrients. Various bacterial species, such as those belonging to the genera *Pseudomonas*, *Bacillus*, *Rhodococcus* or *Acinetobacter*, have shown a high capacity to degrade polycyclic aromatic hydrocarbons, chlorinated solvents, heavy metals, pesticides, and pharmaceutical compounds, thanks to enzymes such as oxidases, peroxidases, and dehydrogenases (Strong & Burgess, 2008; Ghosal et al., 2016). Additionally, another relevant approach is phytoremediation, which uses higher plants to remediate contaminated environments. Plants can absorb heavy metals through roots (phytoextraction), transform them through enzymatic reactions (phytodegradation), leach them from water bodies (rhizofiltration), or even release them in volatile form (phytovolatilization) (Ali et al., 2013; Pilon-Smits, 2005). Species such as *Brassica juncea* (Indian mustard), *Helianthus annuus* (sunflower) and Populus spp. (poplar) have been widely used in remediation studies of soils contaminated by metals such as cadmium, lead, and arsenic, given their rapid growth rate, high biomass, and accumulation capacity. It should be noted that the success of these biological processes is strongly conditioned by the physico-chemical characteristics of the medium, such as pH, humidity, oxygen availability, temperature, and contaminant composition, which can limit remediation efficiency in hostile environments or in the presence of highly recalcitrant compounds (Megharaj et al., 2011).

5.2 *Types of Bioremediation: In Situ, Ex Situ, Passive and Active*

The classification of bioremediation strategies is commonly established based on the place of application and the level of technological intervention required. In this sense, four main categories are recognized: in situ, ex situ, passive, and active.

- In situ bioremediation is carried out directly on the contaminated site, without the need to excavate or transport the affected material. This modality is more economical and less disruptive, and is used, for example, in the biostimulation of hydrocarbon-contaminated soils, where nutrients and oxygen are added to promote the activity of indigenous microorganisms (Vidali, 2001).
- On the other hand, ex situ bioremediation involves the removal of contaminated material to a treatment facility, such as bioreactors, compost piles or biobeds, where conditions can be tightly controlled. This approach allows for greater efficiency and monitoring of the process, but also entails higher logistical and energy costs (Tyagi et al., 2011).
- Passive bioremediation is based on stimulating natural degradation processes without continuous direct intervention. It is used, for example, in constructed wetlands or biobarren areas, where favorable conditions are created for degrading microorganisms to act spontaneously.
- In contrast, active bioremediation involves more intensive manipulation, such as the addition of specific microbial strains (bioaugmentation), pH, temperature or oxygenation adjustments, or the incorporation of surfactants and amendments to facilitate degradation (Singh et al., 2010).
- Each of these strategies should be selected based on the type and concentration of contaminant, the matrix affected (water, soil, sediment), site characteristics, and environmental restoration objectives (Fig. 2.4).

5.3 *Limitations of Traditional Technologies*

Despite its many benefits, bioremediation has inherent limitations that have motivated the development of hybrid approaches and emerging technologies such as nanotechnology. First, many organic and inorganic contaminants are recalcitrant, i.e., they resist biological degradation owing to their complex chemical structure, such as polychlorinated biphenyls (PCBs), perfluorinated compounds, or organochlorine pesticides. Also, the presence of complex mixtures of pollutants, frequent on industrial sites, can inhibit microbial activity and make it difficult to choose suitable organisms for their treatment (Boopathy, 2000). In addition, the efficiency of the process depends on specific environmental conditions: optimal levels of oxygen, temperature, humidity, pH, and nutrient availability. When these conditions are not present, microbial metabolism slows down or stops, affecting degradation kinetics

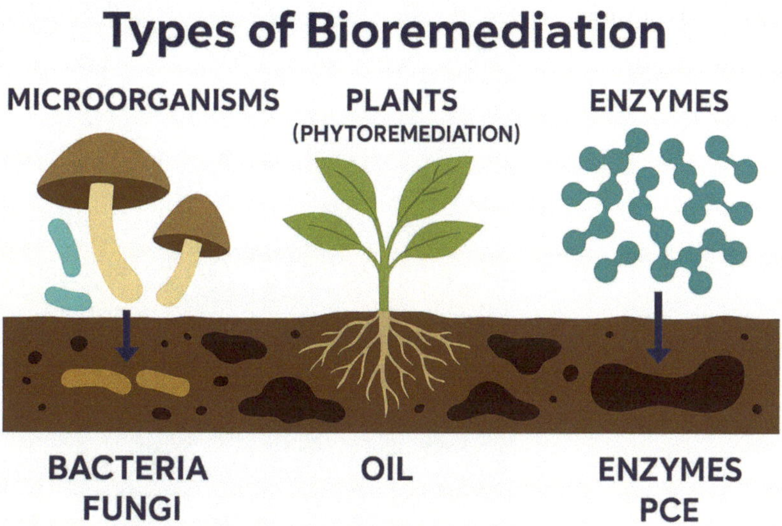

Fig. 2.4 Types of bioremediation. PCE refers to perchloroethylene, on which enzymatic bioremediation acts

(Megharaj et al., 2011). Another challenge is the speed of the process. Bioremediation can take weeks or even months to reach acceptable levels of decontamination, limiting its application in situations requiring rapid remediation. Finally, there is the risk of direct toxicity of contaminants to the microorganisms or plants used, which may compromise the viability of biological processes (Ghosal et al., 2016). In response to these limitations, the scientific community has explored complementary technologies—including the application of nanomaterials—as tools to accelerate bioremediation processes, broaden their range of applicability, and improve their efficiency under adverse conditions.

6 Introduction to Nanomaterials Applied to Remediation

6.1 *Definition and Main Characteristics of Nanomaterials*

Nanomaterials are substances that have at least one of their dimensions on the nanometer scale, i.e., below 100 nm. On this scale, the materials exhibit extraordinary physicochemical properties (Fig. 2.5), that are not manifested in their bulk form, mainly because of their high surface-to-volume ratio, quantum confinement effects, and a higher proportion of atoms on the surface (Morose, 2010; Simeonidis et al., 2016). These properties endow nanomaterials with enhanced chemical reactivity,

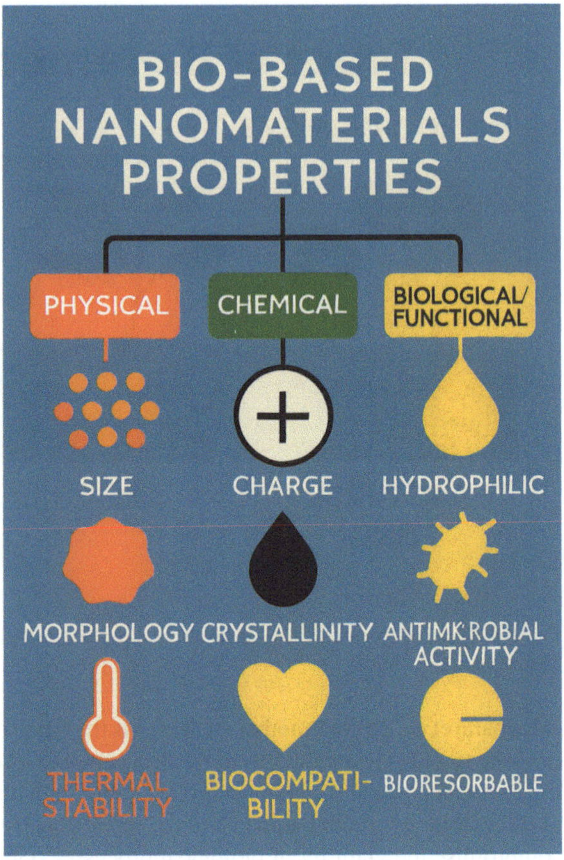

Fig. 2.5 Basic physicochemical properties of bio-based nanomaterials derived from biological sources

increased adsorption capacity, elevated electrical conductivity, increased catalytic activity, and the possibility of surface functionalization, making them highly versatile tools for environmental applications, especially in monitoring and remediation processes of contaminated soils, waters, and sediments (Chen et al., 2020). Among the most frequently used nanomaterials in environmental remediation are:

- Metallic nanoparticles, such as zero valent iron (nZVI), silver (Ag), or gold (Au), known for their redox capacity and bactericidal effects.
- Metal oxides, such as titanium dioxide (TiO_2), zinc oxide (ZnO), and magnetite (Fe_3O_4), used in photocatalytic processes and adsorption.
- Carbon nanotubes (CNTs), which offer a high surface area, thermal stability, and the possibility of chemical modification.
- Quantum dots, with unique optical properties that allow their use in the detection and degradation of contaminants.
- Hybrid nanocomposites, composed of nanomaterials and functional polymers (such as chitosan), which combine adsorption capacity, microbial immobilization, and post-treatment recovery (Kumar et al., 2019; Singh et al., 2022).

6.2 Physicochemical Properties Useful in Environmental Processes

The attractiveness of nanomaterials in the environmental field lies in a set of key properties that enhance their performance against persistent pollutants (Fig. 2.6): First, their high specific surface area allows for greater exposure of active sites, which significantly increases their ability to adsorb metal ions, pesticides, pharmaceutical compounds, or polycyclic aromatic hydrocarbons (Wang & Keller, 2008). Second, their ease of chemical functionalization allows their surfaces to be modified with specific groups (such as amino, thiol, or carboxyl), which favors selectivity toward certain pollutants or compatibility with biological systems (Chen et al., 2020; Zhao et al., 2021). Redox capacity is another distinctive feature, especially in nanoparticles such as nZVI, which can reduce toxic compounds such as hexavalent chromium (Cr(VI)) to its less toxic trivalent form (Cr(III)) by direct reduction mechanisms (Mishra & Pradhan, 2020). Finally, certain nanomaterials such as TiO_2 and ZnO exhibit photocatalytic activity when exposed to ultraviolet or visible light, generating reactive oxygen species (ROS)—such as hydroxyl radicals (−OH)—capable of oxidizing recalcitrant organic compounds such as industrial dyes, phenols, or pesticides (Chen et al., 2020; Yuan et al., 2021).

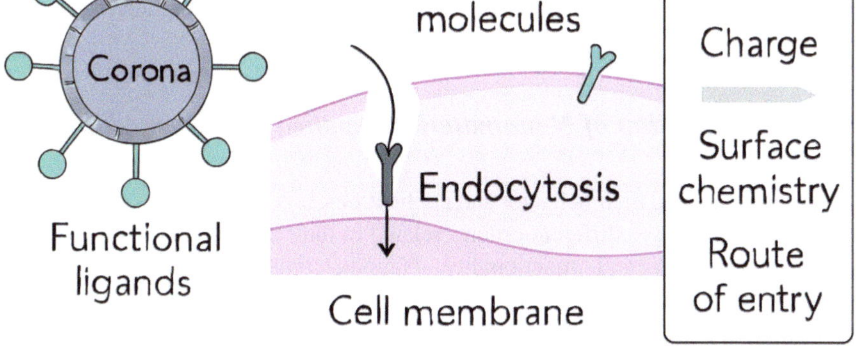

Fig. 2.6 The nanoparticle–cell membrane interface. Size, surface charge/pKa, and ligand chemistry are key physicochemical properties of nanoparticles that shape their interactions with cell membranes and, therefore, determine their cellular entry pathways

6.3 Mechanisms of Action in Bioremediation Processes

Nanomaterials applied in remediation do not act in isolation; on the contrary, their effectiveness is often enhanced when integrated with biological strategies such as bioremediation, giving rise to what is now known as nanobioremediation. In this context, the most relevant mechanisms of action include:

- Adsorption: Many nanomaterials have highly active surfaces that allow the retention of contaminants through physical (Van der Waals forces) or chemical (covalent or ionic bonds) mechanisms. For example, metal oxides and carbon nanotubes have demonstrated high efficiency in adsorbing heavy metals and organochlorine pesticides (Chowdhury et al., 2012).
- Catalysis: Some nanomaterials act as catalysts or photocatalysts, accelerating chemical reactions such as the oxidation of organic pollutants. TiO_2 and ZnO, for example, generate ROS under UV light, which can break resistant molecular bonds, facilitating biodegradation (Zhou et al., 2014).
- Transport of active compounds: Nanoparticles can also be used as delivery vehicles to transport enzymes, microorganisms, or nutrients into hard-to-reach contaminated areas, increasing the bioavailability of these elements in heterogeneous matrices such as clay soils or deep sediments (Wang & Keller, 2008).
- Support for microbial immobilization: By incorporating nanomaterials into polymeric matrices (e.g., chitosan–Fe_3O_4), hybrid systems can be created that protect microorganisms from adverse conditions (extreme pH, presence of heavy metals) and maintain their viability and functionality during the remediation process (Singh et al., 2022; Kumar et al., 2019).

These mechanisms, when combined, have been shown to significantly increase remediation efficiency in highly contaminated and complex environments, whether in soil, groundwater or surface water. In addition, the possibility of developing smart controlled-release systems, integrated sensors, and bio-reactive platforms makes nanomaterials key tools for the future of environmental engineering.

6.4 Classification of Nanomaterials Applied to Bioremediation

To better understand their behavior and applications, it is necessary to classify nanomaterials according to different criteria related to their composition, structure, origin, and environmental functionality. Nanomaterials used in environmental remediation processes can be classified in several ways, depending on the criteria adopted (Table 2.1.). The most relevant classifications are described in the following subsections.

Table 2.1 Types of nanomaterials used in environmental bioremediation, contaminants treated, and mechanisms of action

Type of nanomaterial	Contaminants treated	Main mechanism of action	Main advantages	References
nZVI (Zero Valent Iron)	Cr(VI), TCE, PCBS, nitrates	Chemical reduction, adsorption	High reactivity, low cost, redox capability	Mishra and Pradhan (2020), Wang and Keller (2008)
Metal oxides (TiO_2, ZnO, Fe_3O_4)	Pesticides, dyes, phenolics, heavy metals, etc.	Photocatalysis, adsorption	ROS generation, good stability, magnetism (Fe_3O_4)	Chen et al. (2020), Yuan et al. (2021)
Silver nanoparticles (AgNPs)	Bacteria pathogenic bacteria, pesticides, dyes	Antibacterial, oxidation, adsorption	Antimicrobial properties, high reactivity	Singh et al. (2022), Gade et al. (2008)
Nanotubes Nanotubes (CNTs)	Heavy metals, organic compounds, toxic gases	Adsorption, microbial support, transport	Large surface area, conductivity, functionalization	Chowdhury et al. (2012), Zhou et al. (2020)
Quantum dots (CdS, ZnS, CdSe)	Dyes, pharmaceuticals, heavy metals	Photocatalysis, fluorescence, biosensitizing	Light emission, real- time monitoring	Zhou et al. (2014), Chen et al. (2020)
Hybrid nanocomposites (chitosan-Fe_3O_4)	Metals heavy metals, hydrocarbons	Immobilization microbial, adsorption	Biocompatibility, magnetic recovery	Kumar et al. (2019), Singh et al. (2022)
Nanoclays (montmorillonite, bentonite)	Arsenic, lead, cadmium, pesticides	Adsorption, ion exchange	High exchange capacity, microbial support	Lee et al. (2009), Park et al. (2011)
Functionalized silica nanoparticles	Metals heavy metals, phenolic compounds	Selective adsorption, chelation	High porosity, chemically modifiable	Zhao et al. (2021), Liu et al. (2018)
Nanobubbles (NBs)	Hydrocarbons, nutrients, VOCs	Release of oxygen, advanced oxidation	Increased oxygen availability, microbial metabolic improvement	Kikuchi et al. (2020), Terasaka et al. (2011)

6.4.1 According to Their Chemical Composition

- Metal nanoparticles: Constituted by pure metals such as iron (nZVI), silver (Ag), gold (Au), or copper (Cu), they stand out for their high redox activity and antimicrobial properties. The nZVI is widely used for the reduction of pollutants such as hexavalent chromium or chlorinated compounds (Mishra & Pradhan, 2020).
- Metal oxides: They include titanium dioxide (TiO_2), zinc oxide (ZnO), magnetite (Fe_3O_4), among others. They are known for their photocatalytic activity and their efficiency in the degradation of recalcitrant organic compounds (Chen et al., 2020).

- Carbonaceous nanomaterials: These include carbon nanotubes (CNTs), graphene, and graphene oxide (GO). They possess a large surface area and adsorption capacity, as well as being excellent supports for microorganisms and enzymes (Chowdhury et al., 2012; Zhou et al., 2020).
- Quantum dots: These are semiconductor nanoparticles (such as CdS, CdSe, ZnS), with outstanding optical properties, used in both pollutant degradation and fluorescence monitoring (Zhou et al., 2014).
- Modified nanoclays and silicates: Montmorillonite and bentonite are used for their cation exchange capacity and adsorption of heavy metals, pesticides, and emerging contaminants (Lee et al., 2009).
- Hybrid nanopolymers and nanocomposites: Materials that combine inorganic nanomaterials (such as Fe_3O_4) with natural polymers (chitosan, alginate) to improve their biocompatibility and stability in environmental settings (Kumar et al., 2019).

6.4.2 According to Their Morphological Structure

- Spherical nanoparticles: Of homogeneous size, such as nZVI or AgNPs.
- Nanotubes and nanowires: Cylindrical structures such as CNTs, useful in adsorption and biological support.
- Nanosheets: Flat structures such as graphene or graphene oxides, which allow a high level of surface interaction.
- Nanocapsules and nanobubbles: Vehicles for the controlled release of oxygen, enzymes, or nutrients (Kikuchi et al., 2020).

6.4.3 According to Origin

- Synthetic nanomaterials: produced by physicochemical methods (top–down or bottom–up), such as the chemical reduction of metals or controlled precipitation of metal oxides (Simeonidis et al., 2016).
- Biosynthesized nanomaterials: Generated by green methods, from plant extracts, bacteria, or fungi. These materials present greater biocompatibility and a lower ecotoxicological impact (Ahmed et al., 2016; Singh et al., 2022).

6.4.4 According to Their Functionality in Remediation

- Adsorbents: They capture contaminants by physical–chemical mechanisms (CNTs, oxides, functionalized silica).
- Reducers/oxidizers: Involved in direct redox reactions (nZVI, TiO_2, quantum dots).
- Microbial supports: Improve the viability and activity of microorganisms (hybrid nanocomposites).

- Sensors and tracers: Allow real-time monitoring of the remediation process (quantum dots, nano-photocatalysts).

7 Specific Applications of Nanomaterials in Bioremediation

7.1 Bioremediation of Contaminated Soils

Soils constitute one of the most complex and vulnerable environmental matrices to contamination, owing to their capacity to retain organic and inorganic compounds for long periods of time. The accumulation of heavy metals, petroleum hydrocarbons, pesticides, industrial residues, and recently emerging contaminants represents a significant risk to ecosystem health, agriculture, and human health (FAO, 2015; Nizzetto et al., 2016). As mentioned above, the use of nanomaterials in combination with bioremediation techniques has emerged as a highly promising alternative for the remediation of degraded soils. One of the most frequently studied nanomaterials in contaminated soils is nanoscale zero valent iron (nZVI). These particles have shown great efficiency in reducing hexavalent chromium (Cr(VI)), transforming it into its trivalent form (Cr(III)), which is less toxic and less mobile in soil. Zhang et al. (2011) showed that the addition of nZVI to contaminated soils allowed a decrease of more than 90% of Cr(VI) in less than 24 h, even under non-ideal moisture and pH conditions. The redox reaction between metallic iron and chromate anion (CrO_4^{2-}) is mainly responsible for this effect, which has also been confirmed by Mishra and Pradhan (2020), who highlight the rapid kinetics of this process in their review on nZVI in industrial soils.

Another relevant application is the adsorption and degradation of polycyclic aromatic hydrocarbons (PAHs) in soils contaminated by oil spills. Magnetic iron oxides (Fe_3O_4) and functionalized carbon nanotubes have proven to be excellent adsorbents of compounds such as benzo[a]pyrene and phenanthrene, facilitating their subsequent biodegradation by bacteria of the genus *Pseudomonas* or *Sphingomonas* (Chowdhury et al., 2012; Wang et al., 2021). This synergy between nanoadsorption and biodegradation represents an efficient way of treating sites with mixed contamination. Likewise, hybrid nanocomposites based on natural polymers such as chitosan combined with Fe_3O_4 or ZnO nanoparticles have been employed to immobilize heavy metals such as lead (Pb), cadmium (Cd), and arsenic (As) in contaminated agricultural soils. Kumar et al. (2019) demonstrated that a chitosan–Fe_3O_4 nanocomposite applied to a soil contaminated with Cd reduced its bioavailability by more than 70%, while preserving the microbial structure of the soil, which is not the case with more aggressive traditional chemical agents.

The advantage of these nanomaterials lies not only in their large surface interaction capacity but also in their dual functionality: on the one hand, they act directly on the contaminant, and on the other, they protect or stimulate the activity of native soil-degrading microorganisms (Singh et al., 2022). In soils with extreme pH or salinity conditions, this function is especially valuable. However, it is important to mention that the behavior of nanomaterials in soils depends on factors such as texture, organic matter content, moisture, and pH, which can influence their mobility, stability, and persistence. Therefore, laboratory and microcosm tests are recommended prior to field-scale application (Gottschalk & Nowack, 2011).

Finally, there is a growing interest in the development of green nanomaterials produced by biosynthetic routes or with natural biomolecules to minimize ecotoxicological risks. For example, Ahmed et al. (2016) succeeded in synthesizing silver nanoparticles (AgNPs) using *Azadirachta indica* extracts, with antimicrobial properties useful in soils contaminated with pathogenic microorganisms.

7.2 Bioremediation of Surface and Ground Waters

Surface waters (rivers, lakes, canals) and groundwater (phreatic and deep aquifers) represent a vital source of freshwater for human consumption, irrigation, and industrial use. However, these aquatic matrices are increasingly threatened by persistent contaminants such as heavy metals, volatile organic compounds (VOCs), industrial dyes, pharmaceuticals, pesticides, and excess nutrients, many of which are difficult to remove with conventional treatment technologies (Rodriguez-Narvaez et al., 2017). For such a reason, the application of nanomaterials in aquatic bioremediation has become an innovative strategy for the efficient decontamination of these water sources. These materials not only allow a high interaction with dissolved or suspended pollutants but can also act in synergy with native or introduced microorganisms, favoring biodegradation, immobilization, or chemical transformation processes.

One of the most documented examples is the use of zero valent iron nanoparticles (nZVI) for the remediation of groundwater contaminated with metals and chlorinated solvents. Owing to its small size, nZVI can be injected directly into aquifers, allowing in situ reduction of toxic species such as hexavalent chromium (Cr(VI)), trichloroethylene (TCE), and tetrachloroethylene (PCE) to less toxic or harmless forms (Tratnyek & Johnson, 2006). Liang et al. (2013) demonstrated that a single injection of nZVI into a TCE-contaminated aquifer reduced its concentration by 95% in less than 5 days, with effective radial dispersion over a radius of 5 m.

On the other hand, photocatalytic metal oxides, such as TiO_2 and ZnO, have been widely applied for the degradation of organic pollutants in surface waters. These nanomaterials, when exposed to ultraviolet or visible light, generate ROS that attack the molecular bonds of pesticides, dyes, and pharmaceuticals. In a study by Yuan et al. (2021), it was observed that nitrogen-doped TiO_2 was able to degrade more than 90% of the antibiotic ciprofloxacin in surface water in less than 60 min under simulated solar irradiation. In addition, hybrid nanocomposites, such as chitosan–Fe_3O_4 or silica functionalized with amino or thiol groups, have been used as adsorbent systems and supports for microorganisms. These systems offer advantages such as magnetic recovery, biocompatibility, and selective adsorption capacity, allowing the retention of metal ions such as Pb^{2+}, Cd^{2+}, or As^{3+}, while preserving the activity of microbial strains such as *Bacillus subtilis* or *Pseudomonas putida* (Kumar et al., 2019; Zhao et al., 2021).

Another notable advance is the use of AgNPs and copper oxide nanoparticles (CuO NPs) in the remediation of water contaminated with pathogenic microorganisms. Although their bactericidal effect can be detrimental to beneficial microbiotas if not properly controlled, their spot application has been effective in the treatment of wastewater with a high bacterial load (Singh et al., 2022).

As for groundwater, the main challenge is to ensure the mobility and stability of nanoparticles, avoiding their aggregation and sedimentation. To this end, systems stabilized with natural biopolymers or biosynthetic surfactants have been developed, which improves their transport and reduces their environmental toxicity (Phenrat et al., 2010). Finally, it is worth mentioning the potential of nanobioremediation in integrated systems, such as artificial wetlands (Phenrat et al., 2010). treated with nanomaterials, nanoporous media bioreactors, or nanosensor-assisted bioremediation, which allows real-time monitoring of water quality and treatment efficiency (Chen et al., 2020).

7.3 Bioremediation of Industrial Wastewater

Industrial wastewater represents one of the most critical sources of environmental pollution, owing to the high load of toxic, persistent, and nonbiodegradable compounds it contains. These waters can include heavy metals (such as Hg, Pb, Cd, Cr), organic solvents, synthetic dyes, detergents, phenols, hydrocarbons, and pharmaceutical compounds, which hinder conventional treatment processes and represent a risk to aquatic ecosystems and human health (Rodriguez-Narvaez et al., 2017). Faced with this problem, the combination of bioremediation with nanomaterials has emerged as a highly effective technology to increase the efficiency and speed of decontamination in industrial effluents. Unlike conventional physicochemical treatments, bio-nano hybrid systems allow the simultaneous removal of different types of pollutants through synergistic mechanisms such as adsorption, catalysis, immobilization, and enzymatic degradation.

One of the most frequently studied approaches in the last decade is the use of nanometric metal oxides, such as titanium dioxide (TiO_2) and zinc oxide (ZnO), which act as photocatalysts capable of degrading recalcitrant compounds under ultraviolet or visible irradiation. Chen et al. (2020) demonstrated that the application of nitrogen-doped TiO_2 achieved degradation of more than 95% of the azo dye acid orange 7 in textile industry wastewater in only 90 min, without requiring extreme pH or elevated temperatures. Under similar conditions, ROS generated by ZnO have shown efficacy against phenols and pharmaceutical compounds (Yuan et al., 2021).

On the other hand, chitosan and magnetite (Fe_3O_4) nanocomposites have been successfully applied as adsorbent and microbial support systems in bioreactors for the removal of heavy metals such as Pb^{2+}, Cd^{2+}, and Cr(VI). In a study by Kumar et al. (2019), they were able to reduce the concentration of Pb^{2+} in industrial electroplating effluents by more than 80%, while maintaining the metabolic activity of immobilized bacterial strains of *Pseudomonas aeruginosa* during a 15-day continuous operation cycle.

A significant advance has been the incorporation of nanomaterials in continuous flow anaerobic bioreactors, where the activity of microbial consortia is enhanced by the addition of iron oxide nanoparticles. These particles act as redox mediators that facilitate electron transfer between microorganisms and pollutants, optimizing the degradation of VOCs and metals in industrial wastewater (Zhou et al., 2020). Likewise, silica functionalized with amino and thiol groups has been used to selectively capture heavy metals and industrial dyes in adsorption columns, demonstrating a regeneration capacity exceeding five cycles without significant loss of efficiency (Zhao et al., 2021). This property is crucial for the design of sustainable and economically viable technologies in industries such as chemical, pharmaceutical, and textile. In addition, the successful use of cadmium sulfide quantum dots (CdS QDs) in photobioremediation processes has been reported, acting as nanometer semiconductors that not only degrade pollutants under visible light but also allow monitoring of the kinetics of the process by fluorescence, facilitating the control of parameters in complex industrial systems (Zhou et al., 2014).

Despite these advances, it is important to note that large-scale implementation of these systems still faces technical, regulatory, and economic challenges. The stability of nanoparticles in complex matrices, their post-treatment recovery and the potential toxicity of residual materials are aspects that require careful evaluation before widespread application (Gottschalk & Nowack, 2011). However, the development of green and biocompatible nanomaterials, as well as their integration with automated monitoring systems, provides a glimpse of a new generation of industrial wastewater treatments that combine technological effectiveness, energy efficiency, and environmental sustainability.

8 Ecotoxicological Aspects and Risks of the Use of Nanomaterials

Despite the technological advantages offered by nanomaterials in environmental remediation, their increasing use raises fundamental questions about their environmental safety, interaction with living organisms, and persistence in natural ecosystems. The small scale, high reactivity, and penetration capacity of nanomaterials—properties that make them useful—can also lead to adverse ecotoxicological effects if not properly controlled (Klaine et al., 2008). The main potential risks are discussed below, based on verifiable scientific evidence.

8.1 *Bioaccumulation and Trophic Transfer*

Nanomaterials, because of their size of less than 100 nm, have the ability to cross cell membranes, enter tissues, and accumulate in internal organs of aquatic and terrestrial organisms. This bioaccumulation can trigger direct toxic effects or act as a trophic transfer pathway, affecting higher-level predators (Moore, 2006). Griffitt et al. (2009) observed that AgNPs induce oxidative stress and tissue damage in the gills and liver of zebrafish (*Danio rerio*), even at realistic environmental concentrations. Also, Hardman (2006) pointed out that CdSe and CdS quantum dots, commonly used for their fluorescence, release Cd^2 ions$^+$ upon degradation, which are highly toxic to crustaceans, fish, and mollusks. These findings underscore the need to study not only the acute toxicity of nanomaterials but also their persistence and accumulation in food chains, especially in aquatic environments, where their mobility is greater.

8.2 *Interaction with Native Microbiota*

Soil and water microbiota perform essential ecological functions: nutrient cycling, mineralization, gas regulation, and of course, bioremediation processes. The introduction of nanomaterials can disrupt this microbial balance, affecting key species and reducing the natural efficiency of ecosystems. Rousk et al. (2012) demonstrated that exposure of agricultural soils to CuO NPs significantly reduced bacterial and fungal biomass, altering nitrogen and carbon cycling. Similarly, Ge et al. (2011) reported that the presence of TiO_2 and ZnO nanoparticles in soils decreased bacterial diversity and inhibited key enzymatic activity for pollutant degradation. These alterations not only compromise soil ecosystem services, but can also neutralize or inhibit microorganisms introduced for bioremediation processes, making technological intervention inefficient.

8.3 *Toxicity to Soil and Aquatic Organisms*

Nanotoxicity varies according to the physicochemical properties of the nanomaterial (size, shape, surface area, solubility, coating) and according to the type of organism exposed. In aquatic environments, nanoparticles such as ZnO and TiO_2 have been shown to be cytotoxic, genotoxic, and reproductively harmful to phytoplankton, zooplankton, and fish (Keller et al., 2010). Hardman (2006) also highlighted that nanoparticles used for their electronic functionality—such as quantum dots or doped nanoparticles—can release metal ions or generate ROS that induce cell damage. In soils, Simonin et al. (2016) identified negative effects on microbial activity, metal accumulation in earthworms (*Eisenia fetida*), and reduced germination in sensitive plant species. It should be noted that these effects can be intensified when nanomaterials interact with other contaminants present in the environment, such as pesticides or heavy metals, generating toxic synergies not yet fully understood (Lowry et al., 2012).

8.4 International Regulatory Recommendations

The legal framework for nanomaterials is still limited, although some international organizations have begun to establish guidelines for their characterization, handling, and risk assessment. The Organisation for Economic Co-operation and Development (OECD) has developed a series of guidance documents for the assessment of manufactured nanomaterials, proposing standardized methods for measuring toxicity, mobility, degradation, and exposure (OECD, 2021).

For its part, the US Environmental Protection Agency (EPA) has integrated some nanomaterials into the Toxic Substances Control Act (TSCA), requiring prior information on composition, stability, toxicity, and environmental fate before marketing (EPA, 2020). Important recommendations include:

- To conduct ecotoxicological tests with representative organisms (fish, algae, invertebrates, bacteria).
- To evaluate the persistence and transformation of the nanomaterial under real environmental conditions.
- To include bioaccumulation and mobility parameters in soil and water.
- To promote the use of green and biodegradable nanomaterials, synthesized from natural compounds or through biological routes.

The accelerated development of nanotechnology has outpaced the ability of traditional environmental legislation to adequately assess and regulate the use of nanomaterials. Given their unique nature, emerging properties, and potential persistence in the environment, various international organizations have begun to establish specific standards, guidelines, and regulatory frameworks to minimize ecotoxicological risks and ensure the safe use of these technologies in contexts such as bioremediation, water treatment, and agriculture.

8.4.1 Organisation for Economic Co-operation and Development Standards and Guidelines

The OECD is one of the most active international bodies in the development of guidelines on nanomaterial safety. Since 2006, it has implemented the Manufactured Nanomaterials Safety Program, aimed at promoting international cooperation for risk assessment, standardization of test methods, and regulatory harmonization between countries. Some of its key publications include:

(a) Guidance on Grouping of Nanomaterials and Read-Across Approaches (OECD, 2017), which provides criteria for classifying nanomaterials according to structural similarities, environmental behavior, and toxicity, facilitating their evaluation without the need to duplicate studies.
(b) OECD Test Guidelines for Nanomaterials.

The OECD has adapted many of its test guidelines to include specific nanoscale procedures, including dispersion in biological media, particle size, surface reactivity, solubility, agglomeration, and bioavailability.

8.4.2 Cooperative Evaluation Dossiers (Sponsorship Program)

The dossiers include detailed studies on materials such as nZVI, TiO_2, ZnO, AgNPs, and CNTs, evaluating environmental behavior, toxicokinetics, and toxic effects.

8.4.3 US Environmental Protection Agency Regulations

The EPA has incorporated the control of nanomaterials within the framework of the TSCA, requiring manufacturers and importers to submit detailed information on physicochemical properties, stability, toxicity, and environmental fate before marketing. In 2017, a specific rule for nanomaterials was published under the TSCA, which mandates notification to the federal government on the production, processing, and use of new or existing nanomaterials. The EPA has also funded research projects on eco-toxicity, environmental modeling, and "safe-by-design" nanomaterials.

8.4.4 European Union

Registration, Evaluation, Authorization and Restriction of Chemicals**, and the** European Chemicals Agency**.**

In the European Union, the Registration, Evaluation, Authorization and Restriction of Chemicals (REACH) framework also regulates nanomaterials. Since January 2020, all substances in nano form are required to include specific data in their registrations, such as:

- Particle size and distribution
- Specific surface
- Morphology (spherical, tubular, lamellar)
- Surface reactivity

The European Chemicals Agency (ECHA) provides detailed guidance on REACH compliance for nanomaterials, including testing for toxicity, persistence, and bioaccumulation.

8.4.5 International Organization for Standardization

The International Organization for Standardization (ISO) has issued a number of technical standards focused on the characterization and evaluation of nanomaterials, including:

- ISO/TR 11360:2010—"Nanotechnologies—Methodology for the classification and categorization of nanomaterials."
- ISO/TS 80004 series—Standardized vocabulary for nanomaterials
- ISO 29701:2010—Requirements for assessing the bioavailability and toxicity of nanomaterials under environmental conditions

8.4.6 Future Ethical and Regulatory Considerations

Although there has been notable progress, most of these standards are not yet binding and regulation at the global level remains fragmented. It is recognized that there is an urgent need for:

- Comprehensive legal frameworks to assess the complete nanomaterial life cycle (manufacture, use, disposal).
- Harmonized testing methodologies that consider complex interactions with the environment.
- Adaptive risk assessment, including predictive modeling and alternative approaches such as in silico assessment.
- Incorporation of the perspective of environmental ethics, based on the principle of precaution and sustainability.

9 Conclusion

The integration of nanotechnology in bioremediation has proven to be an effective strategy for environmental decontamination, improving the efficiency of biological processes by optimizing the bioavailability of contaminants and stimulating microbial activity. Recent studies have highlighted the role of metal nanoparticles, photocatalytic oxides, and hybrid nanocomposites in the degradation of persistent pollutants, as well as their application in soil, water, and industrial waste treatment (Sahu et al., 2025; Nadig, 2025). Although nanobioremediation offers significant advantages, its large-scale implementation faces technical and regulatory challenges, particularly in relation to ecotoxicity and mobility of nanoparticles in natural environments. Research has pointed out the need to evaluate the effects of bioaccumulation and trophic transfer of nanomaterials in aquatic and terrestrial ecosystems (Suarez et al., 2025). Therefore, the development of green and biosynthesized nanomaterials that minimize environmental risks without compromising remediation efficiency is promoted. This chapter provides a comprehensive overview of the intersection between nanotechnology and bioremediation, highlighting the need for multidisciplinary approaches to optimize these technologies without compromising the integrity of ecosystems. The future of this area will depend not only on scientific advances but also on the implementation of regulations and environmental management strategies that favor a balance between innovation and sustainability.

References

Ahmed, S., Ahmad, M., Swami, B. L., & Ikram, S. (2016). Green synthesis of silver nanoparticles using Azadirachta indica aqueous leaf extract. *Journal of Radiation Research and Applied Sciences, 9*(1), 1–7. https://doi.org/10.1016/j.jrras.2015.06.006

Ali, H., Khan, E., & Sajad, M. A. (2013). Phytoremediation of heavy metals-concepts and applications. *Chemosphere, 91*(7), 869–881. https://doi.org/10.1016/j.chemosphere.2013.01.075

Aus der Beek, T., Weber, F. A., Bergmann, A., Hickmann, S., Ebert, I., Hein, A., & Küster, A. (2016). Pharmaceuticals in the environment-global occurrences and perspectives. *Environmental Toxicology and Chemistry, 35*(4), 823–835. https://doi.org/10.1002/etc.3339

Boopathy, R. (2000). Factors limiting bioremediation technologies. *Bioresource Technology, 74*(1), 63–67. https://doi.org/10.1016/S0960-8524(99)00144-3

Boxall, A. B., Rudd, M. A., Brooks, B. W., Caldwell, D. J., Choi, K., Hickmann, S., & Van der Kraak, G. (2009). Pharmaceuticals and personal care products in the environment: What are the big questions? *Environmental Health Perspectives, 117*(8), 1155–1162. https://doi.org/10.1289/ehp.0800546

Chen, H., Li, X., Zhu, H., & Wang, Y. (2020). Application of quantum dots in wastewater treatment: A review. *Journal of Cleaner Production, 268*, 122378. https://doi.org/10.1016/j.jclepro.2020.122378

Chowdhury, I., Hong, Y., Honda, R. J., & Walker, S. L. (2012). Mechanisms of multidimensional transport of functionalized carbon nanotubes in porous media. *Environmental Science & Technology, 46*(10), 5488–5494. https://doi.org/10.1021/es204299b

Das, N., & Dash, H. R. (2014). Microbial bioremediation: A potential tool for restoration of contaminated areas. In *Bioremediation of industrial waste for environmental safety*. Springer.

Dzionek, A., Wojcieszyńska, D., & Guzik, U. (2016). Natural carriers in bioremediation: A review. *Electronic Journal of Biotechnology, 23*, 28–36. https://doi.org/10.1016/j.ejbt.2016.07.003

EPA (United States Environmental Protection Agency) (2020). Nanotechnology and the environment. https://www.epa.gov/research/nanotechnology

FAO & ITPS. (2015). Status of the World's Soil Resources (SWSR) – Main Report. Food and Agriculture Organization of the United Nations. https://www.fao.org/3/i5199e/i5199e.pdf

Gade, A., Taylor, M., Spriggs, C., & Gade, P. (2008). Synthesis of silver nanoparticles by fungi and evaluation of their antimicrobial activity. *Biotechnology Advances, 26*(3), 256–263. https://doi.org/10.1016/j.biotechadv.2007.12.002

Gavrilescu, M. (2004). Environmental biotechnology: Achievements, opportunities and challenges. *Dynamic Biochemistry, Process Biotechnology and Molecular Biology, 1*(1), 1–25.

Ge, Y., Schimel, J. P., & Holden, P. A. (2011). Evidence for negative effects of TiO_2 and ZnO nanoparticles on soil bacterial communities. *Environmental Science & Technology, 45*(4), 1659–1664. https://doi.org/10.1021/es103040t

Geissen, V., Mol, H., Klumpp, E., Umlauf, G., Nadal, M., Schuhmacher, M., et al. (2015). Emerging pollutants in the environment: A challenge for water resource management. *International Soil and Water Conservation Research, 3*(1), 57–65. https://doi.org/10.1016/j.iswcr.2015.03.002

Geyer, R., Jambeck, J. R., & Law, K. L. (2017). Production, use, and fate of all plastics ever made. *Science Advances, 3*(7), e1700782.
Ghosal, D., Ghosh, S., Dutta, T. K., & Ahn, Y. (2016). Current state of knowledge in microbial degradation of polycyclic aromatic hydrocarbons (PAHs): A review. *Frontiers in Microbiology, 7*, 1369. https://doi.org/10.3389/fmicb.2016.01369
Gottschalk, F., & Nowack, B. (2011). The release of engineered nanomaterials to the environment. *Journal of Environmental Monitoring, 13*(5), 1145–1155. https://doi.org/10.1039/c0em00547a
Griffitt R. J., Hyndman K., Denslow N. D., & Barber D. S. (2009). Comparison of molecular and histological changes in zebrafish gills exposed to metallic nanoparticles. *Toxicological Sciences 107*, 404–415. https://doi.org/10.1093/toxsci/kfn256
Hardman, R. (2006). A toxicologic review of quantum dots: Toxicity depends on physicochemical and environmental factors. *Environmental Health Perspectives, 114*(2), 165–172. https://doi.org/10.1289/ehp.8284
Keller, A. A., Wang, H., Zhou, D., & Lenihan, H. S. (2010). Stability and aggregation of metal oxide nanoparticles in natural aqueous matrices. *Environmental Science & Technology, 44*(6), 1962–1967. https://doi.org/10.1021/es902987d
Kikuchi, T., Fukushima, T., & Shiratori, S. (2020). Effects of nanobubbles on environmental purification processes. *Environmental Technology Reviews, 9*(1), 12–24. https://doi.org/10.1080/21622515.2020.1720135
Klaine, S. J., Alvarez, P. J., Batley, G. E., Fernandez, I. J., Handy, R. D., Lyon, D. Y., & Lead, J. R. (2008). Nanomaterials in the environment: Behavior, fate, bioavailability, and effects. *Environmental Toxicology and Chemistry, 27*(9), 1825–1851. https://doi.org/10.1897/08-090.1
Kumar, A., Soni, S., & Jhariya, D. (2019). Chitosan-based nanocomposites for remediation of heavy- metal-contaminated water: A review. *Environmental Nanotechnology, Monitoring & Management, 12*, 100267. https://doi.org/10.1016/j.enmm.2019.100267
Kumari, S., Jamwal, R., Singh, D., & otros. (2020). Recent developments in environmental mercury biosorption and bioremediation: A review. *Journal of Hazardous Materials, 387*, 121754. https://doi.org/10.1016/j.jhazmat.2019.121754
Lee, S. M., Tiwari, D., & Lee, Y. J. (2009). Development of organoclay-supported nanoscale zero- valent iron for the removal of Pb(II) from aqueous solutions. *Environmental Engineering Science, 26*(8), 1283–1290. https://doi.org/10.1089/ees.2008.0413
Liang, C., Wang, Z. S., & Bruell, C. J. (2013). Influence of pH on chemical oxidation using permanganate and ferrous ion for TCE degradation. *Chemosphere, 93*(6), 1287–1293. https://doi.org/10.1016/j.chemosphere.2013.06.061
Liu, L., Li, W., Song, W., & Guo, M. (2018). Remediation techniques for heavy metal contaminated soils: principles and applicability. *Science of the Total Environment, 633*, 206–219. https://doi.org/10.1016/j.scitotenv.2018.03.161.
Lowry, G. V., Gregory, K. B., Aiken, G. R., Conradson, S. D., Kester, D. R., Mattasch, T. J., et al. (2012). Transformations of nanomaterials in the environment. *Environmental Science & Technology, 46*(13), 6893–6899. https://doi.org/10.1021/es300839e
Megharaj, M., Ramakrishnan, B., Venkateswarlu, K., Sethunathan, N., & Naidu, R. (2011). Bioremediation approaches for organic pollutants: A critical perspective. *Environment International, 37*(8), 1362–1375. https://doi.org/10.1016/j.envint.2011.06.003
Mishra, S., & Pradhan, D. (2020). Nanoscale zero-valent iron (nZVI) particles for environmental remediation: A review. *Environmental Technology & Innovation, 20*, 101067. https://doi.org/10.1016/j.eti.2020.101067
Moore, M. N. (2006). Do nanoparticles present ecotoxicological risks for the health of the aquatic environment? *Environment International, 32*(8), 967–976. https://doi.org/10.1016/j.envint.2006.06.014
Morose, G. (2010). The five principles of "design for safer nanotechnology". *Journal of Cleaner Production, 18*(3), 285–289. https://doi.org/10.1016/j.jclepro.2009.10.001
Nadig, S. (2025). Nanobioremediation: Microbe-nano solutions to environmental contaminants. *International Journal of Scientific Research in Engineering and Technology, 8*(6).

Nizzetto, L., Futter, M., & Langaas, S. (2016). Are agricultural soils dumps for microplastics of urban origin? *Environmental Science & Technology, 50*(20), 10777–10779. https://doi.org/10.1021/acs.est.6b04140

OECD. (2017). Guidance on grouping of nanomaterials and read-across approaches. https://www.oecd.org/chemicalsafety/nanosafety/publications-nanomaterials.htm

OECD. (2021). Testing Programme of Manufactured Nanomaterials. https://www.oecd.org/chemicalsafety/nanosafety/testing-programme-manufactured-nanomaterials.htm

Park, J. S., Komarneni, S., & Cho, W. J. (2011). Removal of heavy metals using layered double hydroxide. *Clays and Clay Minerals, 59*(4), 313–321. https://doi.org/10.1346/CCMN.2011.0590403

Phenrat, T., Saleh, N., Sarpeshkar, P., Brandt, W., & Lowry, G. V. (2010). Nanoparticle transport in natural soils: Stability and mobility of nanoscale zero-valent iron. *Environmental Science & Technology, 44*(9), 3426–3431. https://doi.org/10.1021/es902974h

Pilon-Smits, E. (2005). Phytoremediation. *Annual Review of Plant Biology, 56*, 15–39. https://doi.org/10.1146/annurev.arplant.56.032604.144214

Rodriguez-Narvaez, O. M., Peralta-Hernández, J. M., Goonetilleke, A., & Bandala, E. R. (2017). Treatment technologies for emerging contaminants in water: A review. *Science of the Total Environment, 576*, 409–432. https://doi.org/10.1016/j.scitotenv.2016.10.110

Rousk, J., Ackermann, K., Curling, S. F., & Jones, D. L. (2012). Impact of silver nanoparticles on microbial communities in soil columns. *Soil Biology and Biochemistry, 49*, 86–95. https://doi.org/10.1016/j.soilbio.2012.01.028

Sahu, A., Manyapu, V., Paliya, S., & Mandpe, A. (2025). Nanotechnology-assisted bioremediation: A multidisciplinary approach to environmental restoration. In *Springer environmental science and engineering* (pp. 403–430). https://doi.org/10.1007/978-3-031-76886-6_17

Simeonidis, K., Stamatis, P., Karalekas, D., & Kalogiannis, T. (2016). Green synthesis of iron-based nanoparticles and their application as adsorbents for wastewater treatment. *Journal of Environmental Chemical Engineering, 4*(1), 147–153. https://doi.org/10.1016/j.jece.2015.11.014

Singh, A., Ward, O. P., & Singh, S. N. (2010). *Applied bioremediation and phytoremediation*. Springer.

Singh, R. P., Tiwari, R., & Shukla, R. (2022). Biopolymer-based nanocomposites for heavy-metal removal from water: A review. *Environmental Research, 203*, 111762. https://doi.org/10.1016/j.envres.2021.111762

Skoogfors L. (2007). Oil spill clean up in Kansas. [Photograph]. Federal Emergency Management Agency (FEMA) Photo Library. https://commons.wikimedia.org/wiki/File:FEMA_-_31024_-_Oil_spill_clean_up_in_Kansas.jpg

Strong, P. J., & Burgess, J. E. (2008). Treatment methods for wine-related and distillery wastewaters: A review. *Bioremediation Journal, 12*(2), 70–87. https://doi.org/10.1080/10889860802052052

Suarez, E. E. A., Roca Jalil, M. E., Fernandez Baldo, M. A., & Cuozzo, S. A. (2025). Nanobiotechnology approaches for the remediation of persistent and emerging organic pollutants. *Environmental Science: Nano, 2*. https://doi.org/10.1039/D4EN00424H

Terasaka, K., Hirabayashi, A., Nishino, T., Fujioka, S., & Kobayashi, D. (2011). Development of a microbubble aerator for wastewater treatment using aerobic activated sludge. *Chemical Engineering Science, 66*(14), 3172–3179. https://doi.org/10.1016/j.ces.2011.03.018

Tratnyek, P. G., & Johnson, R. L. (2006). Nanotechnologies for environmental cleanup. *Nano Today, 1*(2), 44–48. https://doi.org/10.1016/S1748-0132(06)70048-3

Tyagi, M., da Fonseca, M. M. R., & de Carvalho, C. C. C. R. (2011). Bioaugmentation and biostimulation strategies to improve the effectiveness of bioremediation processes. *Biodegradation, 22*(2), 231–241. https://doi.org/10.1007/s10532-010-9394-4

Vidali, M. (2001). Bioremediation: An overview. *Pure and Applied Chemistry, 73*(7), 1163–1172. https://doi.org/10.1351/pac200173071163

Wang, S., & Keller, A. A. (2008). Natural and engineered nanomaterials for environmental applications. *Journal of Environmental Quality, 37*(6), 1796–1808. https://doi.org/10.2134/jeq2008.0066

Wang, Z., Wu, H., & Xu, Y. (2021). Degradation of phenanthrene in soil using carbon-nanotube-supported bacterial consortia. *Environmental Pollution, 274*, 116588. https://doi.org/10.1016/j.envpol.2021.116588

Wikimedia Commons. (2025). Biorremediación: Tratamiento de aguas de un canal contaminado [Fotografía]. Retrieved June 15, 2025 from: https://commons.wikimedia.org/wiki/File:Biorremediacion_Tratamiento_de_aguas_de_un_canal_contaminado.JPG

Wilkinson, J. L., Boxall, A. B. A., Kolpin, D. W., Leung, K. M. Y., Lai, R. W. S., Galbán-Malagón, C., et al. (2022). Pharmaceutical pollution of the world's rivers. *Proceedings of the National Academy of Sciences, 119*(8), e2113947119. https://doi.org/10.1073/pnas.2113947119

Wright, S. L., & Kelly, F. J. (2017). Plastic and human health: A micro issue? *Environmental Science & Technology, 51*(12), 6634–6647. https://doi.org/10.1021/acs.est.7b00423

Yuan, X., Wang, Y., Zeng, G., & Chen, X. (2021). Nanomaterials in the environment and their applications in pollution remediation: An overview. *Environmental Science and Pollution Research, 28*(17), 21038–21058. https://doi.org/10.1007/s11356-020-11465-1

Zhang, W. X., Wang, C. B., & Lien, H. L. (2011). Treatment of chlorinated organic pollutants with nanoscale bimetallic particles. *Catalysis Today, 40*(4), 387–395. https://doi.org/10.1016/S0920-5861(98)00414-3

Zhang, W., Wang, Y., Lee, S. A., & Liu, Z. (2019). Application of nanotechnology in remediation of contaminated soil and groundwater: A review. *Environmental Science and Pollution Research, 26*(12), 12134–12143. https://doi.org/10.1007/s11356-019-04524-y

Zhao, Y., Wang, Y., & Wei, Z. (2021). Functionalized silica nanoparticles for wastewater treatment. *Journal of Environmental Chemical Engineering, 9*(5), 105264. https://doi.org/10.1016/j.jece.2021.105264

Zhou, W., Gu, Y., Li, C., & Qian, X. (2014). Quantum dots as light harvesters in photo-bioreactors for improved wastewater treatment. *Bioresource Technology, 170*, 214–221. https://doi.org/10.1016/j.biortech.2014.07.107

Zhou, M., Wang, H., Li, Y., & Xia, Y. (2020). Enhancing anaerobic digestion of wastewater via conductive nanomaterials: Mechanisms and perspectives. *Water Research, 174*, 115634. https://doi.org/10.1016/j.watres.2020.115634

Chapter 3
Nanosensoring and Restoration of Water Pollution Using Nanomaterials

Magda E. Hernández García, Aldo F. Bazaldúa Rodríguez, and Joel H. Elizondo Luevano

1 Introduction

The availability of natural resources such as water and soil is crucial for the sustainable development of society, as it ensures the proper utilization of ecosystem services. This, in turn, reduces the economic and energy burden associated with securing potable water and fertile soils, both essential for food production and human survival (ONU, 2023). (*SDG-6-Summary-Progress-Update-2021_Version-July-2021_SP*, n.d.).

However, rapid population growth, combined with increasing industrial, technological, and health care demands, has led to alarming levels of pollution in the air, water bodies, and agricultural soils. Major contaminants include carbon monoxide, heavy metals (such as arsenic, chromium, lead, cadmium, mercury, and zinc), hydrocarbons, nitrogen oxides, volatile organic compounds, sulfur dioxide, pesticides, herbicides, fertilizers, and pharmaceuticals, particularly antibiotics, anticancer agents, and endocrine treatments (Abdallat et al., 2022; Yunus et al., 2012).

The presence of these pollutants in ecosystems exacerbates the environmental footprint of modern society and poses significant health risks. These substances can enter the food chain through contaminated water sources and poor management of

M. E. Hernández García
Laboratorio de Parasitología y Farmacognosia, Centro de Investigación Biomédica del Noreste, Nuevo León, Mexico

A. F. Bazaldúa Rodríguez (✉)
Laboratorio de Fitoquímica, Facultad de Ciencias Biológicas, Universidad Autónoma de Nuevo León, Nuevo León, Mexico
e-mail: aldo.bazalduarg@uanl.edu.mx

J. H. Elizondo Luevano
Laboratorio de Ciencias Naturales, Facultad de Agronomía, Universidad Autónoma de Nuevo León, Nuevo León, Mexico

M. S. Heya (ed.), *Natural Alternatives and Nanotechnology Applied to the One Health Approach*, https://doi.org/10.1007/978-3-032-03675-9_3

agrochemicals, ultimately affecting raw and processed food products (Zwolak et al., 2019). This situation is especially critical in developing countries, where environmental regulations are often weak or non-existent.

Scientific research has consistently shown that exposure to environmental pollutants can lead to inflammatory diseases, promote atherosclerosis, alter cardiovascular function, and increase mortality in individuals with chronic conditions such as diabetes mellitus and respiratory diseases (Vaseashta et al., 2007).

These challenges highlight the urgent need for early detection and remediation of water and soil contamination. In this regard, the development of advanced technologies capable of monitoring, detecting, and restoring polluted environments, particularly rivers, lakes, aquifers, and agricultural soils, has become imperative. Nanotechnology offers a wide array of innovative solutions to address these challenges (Yunus et al., 2012).

The application of nanomaterials such as carbon nanotubes, zeolites, polymeric nanocapsules, nanoparticles, biopolymers, nanofilters, nanoclays, dendrimers, nanosensors, and photocatalytic nanoparticles has proven effective in the remediation of contaminated environments (Grozdanov et al., 2023; Saldivar & Walsh, 2015). These materials can be tailored to various approaches thanks to their unique properties, including high reactivity, large surface area, and enhanced removal capacity, making them valuable tools for the ecological restoration of degraded water and soil systems.

2 Identification of Water Contaminants

Access to water in optimal conditions not only provides direct health benefits but also significantly impacts the economic development of a region. Water contamination of both surface and groundwater bodies is closely related to the level of social and economic development, as well as the presence and enforcement of environmental regulations in each country.

Water pollution occurs when waste and harmful substances are introduced into aquifers, posing a major public health issue. The main sources of water contamination include industries that indiscriminately discharge toxic waste into canals, streams, or rivers, which may eventually flow into the ocean, disrupting aquatic ecosystems. In addition, spills from ships, processing plants, and offshore oil platforms cause environmental disasters that negatively affect coastal communities, commerce, and tourism (de las Grijalva Endara et al., 2020).

Common freshwater contaminants include pathogenic microorganisms such as enteric bacteria, protozoa, and cyanobacteria, many of which serve as indicators of fecal contamination. These pathogens are responsible for waterborne diseases and can trigger gastrointestinal outbreaks when contaminated water is consumed (Noriega-Treviño et al., 2012; de Ramos Mancheno, 2024; Ríos-Tobón et al., 2017). Therefore, disinfection and appropriate treatment of water are essential to ensure its safety for human consumption.

Additionally, chemical contaminants mainly originate from industrial, hospital, and agricultural activities. These include plasticizers such as bisphenol-A, flame retardants, fuel additives, analgesics, anesthetics, pharmaceuticals for cardiovascular diseases and cancer, endocrine-disrupting compounds, and antibiotics. Among the most frequently detected antibiotics in contaminated water are ciprofloxacin, levofloxacin, ofloxacin, oxytetracycline, and doxycycline, which represent a serious concern owing to their role in promoting antimicrobial resistance. Furthermore, pesticides and fertilizers, especially soluble nitrates and phosphates, are often discharged into wastewater and leach into aquifers, leading to eutrophication, deterioration of water bodies, disease transmission, antibiotics resistance, and contamination of both soil and food (Abdallat et al., 2022; Hussain et al., 2017).

In light of these challenges, nanotechnology provides a promising and innovative solution for detecting and removing a wide variety of water pollutants, including heavy metals, dyes, and pharmaceutical residues. Nanosensors offer remarkable advantages for real-time water-quality monitoring, being capable of detecting a broad range of contaminants, even at trace levels, owing to their high degrees of sensitivity and specificity (Adam & Dhahi, 2022).

3 Sensors and Materials for Contaminant Detection

Currently available nanosensors on the market, although still limited, as noted in the literature by RapidChek and Watersafe along with many experimental sensors, are classified based on the type of nanomaterial employed. The type of material not only determines the sensor's structural characteristics but also dictates the range of analytes that can be detected, including heavy metals, microorganisms such as protozoa, bacteria, and viruses, as well as pharmaceuticals and agrochemicals (Vikesland, 2018).

3.1 Colorimetric Sensors

3.1.1 Detection of Heavy Metals

Heavy metals are a group of chemical elements with high atomic densities that exhibit significant toxicity to living organisms in aquatic ecosystems. Their toxicity is primarily due to their persistence in the environment and capacity for bioaccumulation, which can lead to severe health issues in humans, ranging from organ damage to death.

Elements such as lead, chromium, cadmium, mercury, arsenic, and copper are among the most commonly monitored. However, because of differences in chemical behavior, their detection in water requires distinct monitoring strategies, leading to the development of a variety of nanosensor technologies.

Colorimetric Sensors Based on Gold and Silver Nanoparticles

In recent years, colorimetric sensors based on gold (AuNPs) or silver nanoparticles (AgNPs) have gained significant attention owing to their advantageous properties. AuNPs are considered among the most stable nanomaterials because of their chemical inertness and resistance to oxidation. In contrast, AgNPs, although slightly less stable, offer remarkable features such as stronger plasmonic excitation, higher molar extinction coefficients than AuNPs of the same size, superior light interaction compared with any known organic or inorganic chromophore, and most importantly, lower production costs.

To function effectively as sensors, these nanoparticles must exhibit uniformity in both size and shape, with spherical morphology being the most common. Nanoparticles can be synthesized through chemical, physical, or biological methods (Asif et al., 2022; Hammami et al., 2021; Hemlata et al., 2020; Kedar et al., 2022), although chemical synthesis remains the most widely used. This approach typically involves the reduction of metal salts to generate nanoparticle suspensions, often with the aid of organic solvents to achieve uniform particle morphology and size.

Once synthesized, the nanoparticles are coated with stabilizing agents to prevent aggregation and excessive growth. These capping agents chemically bind to the nanoparticle surface, allowing for subsequent interaction with target metal ions or analytes. The selectivity of the sensor toward specific metal ions depends on the structural features of the capping agent. Moreover, the specificity and sensitivity of the detection can be enhanced by functionalizing the surface with additional moieties such as sugars or amino acids (Sener, 2014; Sugunan et al., 2005), as shown in Fig. 3.1.

The ability of these AuNPs or AgNPs to act as eco-friendly sensors for heavy metals lies in their aggregation behavior upon exposure to metal ions. This interaction induces a visible color change in the nanoparticle solution, allowing for simple, equipment-free detection of heavy metals in aqueous environments.

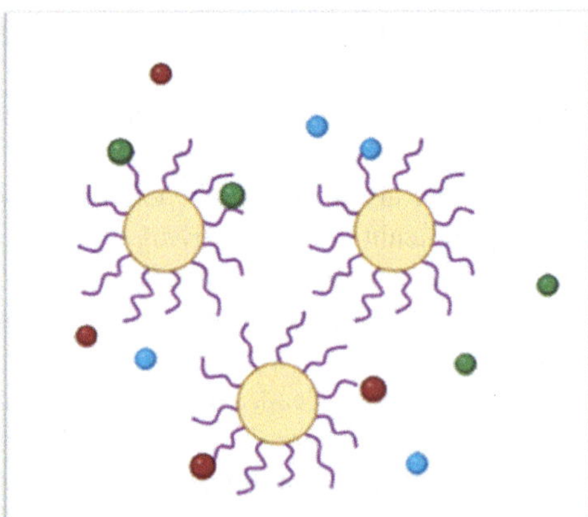

Fig. 3.1 Representation of the binding between gold and silver nanoparticles with heavy metal ions, which attach specifically according to their respective protective capping agents and any added modifiers

Fluorescence Sensor: Carbon Quantum Dots

The development of this type of material as a contaminant sensor was initially accidental. Carbon quantum dots (CQDs) were first identified by (Xu et al., 2004) during the purification of single-walled carbon nanotubes, from which they obtained particles smaller than 10 nm. These particles were later described and characterized, leading to the development of specific synthesis methods, known as top–down and bottom–up approaches. The main difference between these two methods is in how CQDs are generated.

In the top–down method, larger carbon materials are broken down using techniques such as laser ablation, arc discharge, and chemical or electrochemical oxidation. On the other hand, the bottom–up approach synthesizes CQDs from chemical precursors using pyrolysis, carbonization, hydrothermal processes, or microwave-assisted synthesis. Following the classical development of these techniques, alternative synthesis routes were explored using natural materials such as fennel seeds, chitosan, egg-yolk oil, orange juice, lemon peel, bee pollen, collagen, humic substances, hair, peanut shell, soy milk, cashew gum, and garlic. This allowed for a more sustainable and environmentally friendly synthesis method without compromising the homogeneity of the CQDs (Dager et al., 2019; Devi et al., 2019; Long et al., 2021).

The optical properties of CQDs have been explained through several theories, such as quantum confinement or surface-related components. In these scenarios, photoexcitation of charge carriers in CQDs occurs, involving the transition of electrons from the valence band to the conduction band. This generates electron-hole pairs, which allow for the identification of metal ions through intermolecular interactions with the surface functional groups of CQDs—mainly hydroxyl, carbonyl, and carboxyl groups. These groups act as emissive traps for the excited electrons, enabling selective detection (Devi et al., 2019; Yan et al., 2019; Zhu et al., 2015). The effectiveness of CQDs increases when post-synthesis modifications introduce additional functional groups on their surface, thereby improving their sensitivity and selectivity compared with pristine CQDs. These modifications facilitate the detection of a broader range of heavy metals, even at very low concentrations. Thus, both pristine and modified CQDs are capable of detecting ions such as mercury, selenium, iron, copper, chromium, and arsenic. When compared with classical analytical techniques used for detecting such contaminants in water, CQDs offer several advantages: they are fast, simple, and efficient, making them an ideal alternative for monitoring the rapid increase in water contamination by heavy metals.

3.1.2 Detection of Pharmaceuticals as Water Pollutants

The accelerated population growth in many regions of the world has led to overcrowding in urban areas and challenges in housing distribution. A logical consequence of this is the strain on sanitation infrastructure, which increases the likelihood of direct contamination of water bodies such as rivers, lakes, and aquifers. One of the common components of wastewater is the presence of pharmaceuticals. These compounds are excreted through urine and feces by humans and domestic animals, as they are not fully metabolized in the body.

Moreover, industrial activities, hospitals, and cultural habits also contribute to pharmaceutical pollution. Improper disposal of unused or expired medications—often flushed down sinks and toilets—results in their accumulation in leachates, landfills, sludge, and wastewater treatment plants, where they ultimately contaminate aquatic ecosystems (Hussain et al., 2017; Kayode-Afolayan et al., 2022).

This alarming situation underscores the urgent need to develop highly sensitive technologies capable of detecting water contamination at low concentrations, thereby enabling the timely mitigation of ecological damage or public health risks, such as the emergence of antibiotic-resistant bacteria.

Copper Nanoparticles

The identification of drugs in various types of samples, such as water, secretions, or food, is normally identified using chromatographic techniques such as high-performance liquid chromatography (HPLC) or gas chromatography coupled with mass spectrometry (GC-MS), as well as some physical and biological methods. Although these techniques have given excellent results for many years, their application is limited with respect to sample preparation and the detection limit that these chromatographic techniques have, in addition to not being able to be used in the area of contamination damage (Nzilu et al., 2023).

One of the ideal options for solving the problem of identifying in situ the problem of drug contamination is the use of AuNPs and AgNPs, obtaining excellent results. However, the cost involved is very high compared with copper, which shares many of the optical properties, size, and application with AuNPs and AgNPs, while exhibiting unique properties such as reactivity. This makes copper a suitable option as a nanosensor for identifying drug contamination in water.

The synthesis of copper nanoparticles (CuNPs) shares the same similarity with other particles of metallic origin, through chemical, physical, and biological methods. Although the methods mentioned above are somewhat equivalent, the synthesis of nanoparticles used as nanosensors prioritizes chemical synthesis, in which copper ions are used in a solution with a reducing agent such as sodium borohydrate or ascorbic acid, with a capping agent of polyvinylpyrrolidone.

The biological alternative in the mechanism of synthesis of CuNPs emerges as an option that represents a lower impact on the ecosystem. It is here that the synthesis of CuNPs using plant extracts stands out, presenting a high composition in families of natural products that have hydroxyl groups (–OH) such as flavonoids, lignans, or other similar ones, allowing this group to reduce the copper ions stabilizing the nanoparticle. Just as plant extracts allow this type of synthesis, microorganisms also play an interesting role where bacteria of the genus *Pseudomonas* reduce copper ions in the culture medium, achieving the same results as plant extracts (Antonio-Pérez et al., 2023; Shantkriti & Rani, 2014; Tejaswi & Devi, 2025).

The use of CuNPs as a sensor to identify antibiotics and other compounds lies mainly in their ability to interact with the functional groups of the capping agents, thus facilitating the identification of antibiotics such as ceftriaxone, cefazolin, cefoperazone, rifampicin, among others. There are important variations in the type of interaction with different drugs if the capping agents are modified by a different material such as polyethylene glycol, pillararene, polyvinyl alcohol, sodium citrate, and even biopolymers (Javed et al., 2020; Teng et al., 2025), resulting in some of these variations of capping agents being suitable substrates for Raman spectroscopy, which allows the efficiency of CuNPs to be increased in their role as nanosensors (Markina et al., 2020; Markina et al., 2020).

3.2 *Chemoresistor Sensors: Nanowires and Nanotubes*

3.2.1 Graphene-Based Chemoresistors

Other types of sensors used for the detection of chemical and biological analytes include graphene-based chemoresistors. These sensors are highly sensitive and selective, capable of detecting simple or complex analytes, and are effective in identifying a wide range of contaminants, including organic compounds, heavy metals, and pathogenic microorganisms, based on specific recognition elements. Chemoresistors are devices with simple structures composed of two electrical contacts connected by a solid material such as graphene. They convert a chemical signal into an electrical signal depending on the surface interaction between the sensors and the analytes (Shin et al., 2025; Zubiarrain-Laserna & Kruse, 2020).

3.2.2 Nanowires

Nanowires are nanostructures with high sensitivity and precision in detecting biological molecules. They are widely used in areas such as disease diagnostics in medicine, food safety, and environmental monitoring (Dafhalla et al., 2024). The use of nanowire in the decontamination of wastewater can also generate contamination, so it is important to identify the types of waste and thus correctly choose the type of nanowire to be used. Nanowires have different types of conductive properties: metallic, semiconductor, and insulating; their electrical conductivity depends on their diameter, length, and the material from which they are made. The surface of nanowires has a high level of interaction with other materials, which allows compounds or microorganisms to adhere (Alshaybani et al., 2024).

There are gold nanowires that represent a signal transducer that recognizes indicators such as heavy metals, metal oxides, polymers, and other compounds (He et al., 2020). These one-dimensional nanostructures that present high surface-to-volume ratio and have the capacity of self-assembly can be classified into two types, monocrystalline with a diameter of less than 2 nm and polycrystalline with a diameter greater than 5 nm, are synthesized by atomic-cluster assembly, and show good mechanical strength. However, the environment affects the structure, and they lose stability when the outer layer is removed.

Some nanowires are made with a polymeric material, which coats the surface of the membrane or the polymer is dispersed during the elaboration of the membrane. These membranes help to improve the performance of its permeability, selectivity, and hydrophilicity, which allows the accumulation of contaminants to be reduced, this considering the type of polymer with which these membranes are made; therefore, the compatibility between the nanomaterial and the polymer must be reviewed to generate effective membranes (Castro-Muñoz & Castro-Muñoz, 2020).

Polymeric membranes are the most widely used in contaminant separation processes, owing to their high performance and low cost; however, the disadvantages of working with polymeric membranes are that they show little tolerance to high temperatures, corrosives, and organic solvents. The use of various nanomaterials has been considered to make modifications to the membranes, because each type of membrane shows selectivity, permeability, stability, and high processing costs; thus, modifications have been made to the surface of the membrane using other materials such as nanoparticles, two-dimensional layer materials, and nanofibers, which improves the structures, generating better performance and application in wastewater treatment (Khraisheh et al., 2021).

Inorganic membranes are also used in the remediation of water contaminated by hydrocarbons, for which it has been shown that incorporating silica nanoparticles into an inorganic membrane generates a hydrophilic modification of the membrane, improving its performance. Porous ceramic membranes are used to separate water–oil emulsions, as they present mechanical resistance; however, it is difficult to achieve membranes with small pores and antifouling properties (Khraisheh et al., 2021; Wang & Yu, 2013).

Metal-Oxide Nanowires for Detection of Organic Compounds

For the detection of organic compounds, metal-oxide nanowires are used, as they show a high ratio between surface area, volume, and the ability to interact with specific compounds. Among the metal-based nanoparticles are titanium dioxide, zinc oxide, copper oxide, and aluminum oxide, which are used to eliminate dyes present in water, some pharmaceutical compounds, aromatic hydrocarbons, among others. Metal-oxide nanowires detect organic molecules through changes in electrical conductivity owing to the interaction of the nanowire and the molecule, affect the electrical resistance and generate a signal indicating the presence and concentration of the compound. The use of these nanowires allows for faster, more efficient, and low-cost remediation, helping to reduce the potential for residual toxicity (Dabas, 2025; Naseem & Durrani, 2021).

Metal oxides have several functional characteristics that are given by their structure, composition, and morphology, which allows them to be used as adsorbents of wastewater pollutants and to purify industrial effluents through the attraction between the surface of the nanowire and the molecule to be removed. This occurs through electroweak interactions such as dipole–dipole forces, London forces, and Van der Waals interaction. Titanium-oxide nanowires are stable and do not generate side effects; many are resistant to acidic and alkaline media, have a strong adsorbent activity, and have high concentrations of hydroxyl groups on their surface, which allows contaminants to be absorbed. On the other hand, copper-oxide nanowires are used to remove heavy-metal ions and toxic dyes and zinc-oxide nanowires are very effective in removing anionic organic compounds; this is because of electrostatic attraction with anionic particles (Alhalili, 2023; Ghosh et al., 2022).

3.2.3 Nanotubes

One-dimensional nanostructures composed of carbon, known as carbon nanotubes (CNTs), are classified into single-walled (SWCNTs) and multi-walled (MWCNTs) nanotubes. The main differences lie in their diameter and the number of graphite layers. SWCNTs typically have a diameter of about 1 nm and consist of long graphite sheets, whereas MWCNTs are composed of multiple rolled graphite layers forming tubes of different diameters. This structural variability grants them distinct properties and enables their application in water purification—by removing or inactivating pathogens, adsorbing heavy-metal contaminants, and transforming hazardous compounds into less harmful ones. The ends of CNTs can be open or closed, often capped with fullerene-like structures (Arora & Attri, 2020; Maubert et al., 2009).

Carbon nanotubes are cylindrical molecules made of a tubular structure, where carbon atoms are arranged in a hexagonal network. They exhibit a unique combination of properties depending on their structure, wall number, and chemical chirality. However, their physical surface properties such as solubility and dispersibility can be modified to enhance interactions with other molecules (Lariza et al., 2012). The orientation of the hexagonal lattice with respect to the longitudinal axis of a nanotube defines an important structural property called chirality. Chirality determines the rolling direction and chiral angle of the nanotube, which in turn influence its electrical properties (Zaporotskova et al., 2016).

Traditionally, activated carbon adsorption has been used to remove metals from water. Nowadays, single- or multi-walled CNTs are favored owing to their specific features, such as high chemical stability and large surface area. These properties make them suitable for treating wastewater containing metal ions (Gelover Santiago, 2015).

Many organic contaminants in water are adsorbed on both the outer layer and the inner surface of nanotubes, thanks to the functional groups exposed on their surface. However, to determine the absorption rate of CNTs, parameters such as surface area, pore density, nanotube functionality, and purity must be considered. These factors influence the adsorption sites, which include interstitial channels, the inner region, grooves, and the outer surface. Carbon nanotubes also exhibit superior water permeability, electrical conductivity, and great mechanical strength, which contribute to enhanced water desalination capabilities (Arora & Attri, 2020).

Carbon nanotubes are also used to treat water contaminated with pesticides, which are classified according to their chemical composition as organochlorines, organophosphates, carbamates, and pyrethroids, all of which are harmful to the environment and living organisms. Organochlorine compounds are bio-accumulative environmental pollutants used for pest control; these compounds are toxic and carcinogenic, and affect the nervous, immune, and reproductive systems. Organophosphates are highly toxic, they are mainly used in agriculture to control or eliminate pests, and are also found in the manufacture of plastics and solvents; poisoning by these pesticides affects the functions of the nervous system. Carbamates are used in pest control in various crops; they are fat-soluble, toxic, genotoxic, and carcinogenic compounds, and cause respiratory and neurological damage (Alghuthaymi et al., 2020; Firozjaee et al., 2018; Vazquez-Duhalt & Vazquez-Duhalt, 2015). Carbon nanotubes are a reliable option to remove these pesticides, owing to their high sensitivity in detecting them; they present a larger surface area and greater adsorption capacity, which allows a large amount of compounds that are present in contaminated water to be removed.

Carbon nanotubes are of great utility in removing heavy metals such as zinc, and metal ions of cadmium, copper, nickel, lead, among other contaminants, such as hydrocarbons and microorganisms (Das et al., 2014).

3.3 Electrochemical Sensors

Electrochemical sensors are used as innovative methods for the electroanalysis of pesticides, pharmaceuticals, and other substances. Their operation is based on oxidation and reduction reactions that occur during the interaction with other molecules, where the interaction between an analyte and an electrode result in a variation of an electrical signal (Zaporotskova et al., 2016). Typically, these sensors consist of three electrodes: a reference electrode that maintains a stable potential, a counter electrode related to the solution, and a working electrode. These sensors serve as diagnostic and detection tools for many substances, including contaminants, through electrochemical analysis by recording changes in electron transfer occurring at the electrode due to reactions such as variations in electrode resistance, adsorption of electroactive compounds, or charge transfer at the electrode surface (Choudhury, 2020; Liu et al., 2022).

Electrochemical sensors are also employed in water-quality analysis, monitoring nitrate levels, detecting phosphates via cobalt-based electrodes, and identifying heavy-metal ions. Currently, these sensors demonstrate stability, reproducibility, and a high degree of sensitivity and selectivity for detecting water contaminants that pose a global problem (Kanoun et al., 2021; McGarrity & Zhao, 2023).

Electrochemical sensors represent a viable solution in the treatment of water contaminated with pharmaceuticals; however, there are several types of nanomaterials used for the development of these sensors such as carbon nanomaterials, metal nanoparticles, aptamers. There are some limitations of electrochemical sensors, among them, the useful life, increased sensitivity to interactions with various analytes, and all this due to salinity, pH fluctuations, variations in concentration, and interference caused by organic or inorganic compounds (Bounegru et al., 2025).

4 Water Pollution Remediation

4.1 Heavy Metals

Heavy metals are also water pollutants that originate mainly from discharges of products from various activities, especially industrial processes, and represent a serious global problem. These metals are toxic even at low concentrations. The most common heavy metals contaminating water include cadmium, aluminum, cobalt, copper, iron, and lead. These metals cause various environmental and human health damage. For example, cadmium causes kidney damage, copper induces liver damage and Wilson's disease, and lead causes digestive problems, brain damage, and kidney disorders. Because of these harmful effects, it is crucial to remove heavy metals present in wastewater using methods such as precipitation, solvent extraction, ion exchange, reverse osmosis, and filtration, among others. Recently, nanomaterials have been increasingly used for the detection and removal of contaminant metal ions from water. These include biosorbents, magnetic nanoparticles, activated carbon, graphene, CNTs, and various polymeric materials (Kolluru et al., 2021; Pabón et al., 2020).

Other nanoparticle-based options for heavy-metal removal in water involve noncarbon-based materials such as polymer membranes and iron-oxide magnetite nanoparticles. Metal-oxide nanoparticles are effective in removing toxic metal ions and other contaminants smaller than 300 nm, making them a good option for wastewater treatment. The main metal-oxide semiconductors can remove heavy metals suspended in water to purify it (Baby et al., 2022). However, these materials are expensive and require high temperatures for their synthesis (Kausar & Ahmad, 2024).

4.2 Photocatalysis

Remediation of water contamination by pharmaceuticals, volatile organic compounds, and pesticides. Water contamination by organic compounds such as pharmaceuticals, volatile organic compounds (VOCs), and agrochemicals represents a serious threat to the safe consumption of drinking water for the population. The appearance of these types of pollutants in water bodies only comes from a single origin, an anthropogenic origin that is dictated, as mentioned above, by the population's need for health, food, and technological development.

The alternatives used to solve this type of contamination, such as filtration, the addition of flocculants, distillation, and bioremediation, offer adequate results; however, the possibility of adequately eliminating the contaminants or controlling the toxic by-products that may be generated in the remediation process represents a difficult challenge to overcome.

Photocatalysis represented a novel alternative a few years ago, in which organic compounds with a high molecular weight can be degraded to simpler compounds that represent a lower risk for both the population and the ecosystem. This was achieved by using nanomaterials such as semiconducting metal oxides, for example, TiO_2, ZnO, Fe_2O_3, ZnS, and CdS, known as photocatalysts, which can accelerate the reaction rate without suffering any change. The mechanism of photocatalysis activity was proposed in 2003 by Pareek and Adesina, in which they describe a series of reactions in which different ions are generated, starting with the photoholes and electrons, followed by the aqueous dissociation of the pollutant, then an electrophilic absorption of dissolved oxygen, followed by the protonation of the superoxide anion to finally generate the formation of simpler chemical intermediates, mineral acids, and hydroxylated products (Fig. 3.2) (Baruah & Dutta, 2009; Yunus et al., 2012).

Commonly, for the reduction of pharmaceuticals, VOCs, and pesticides, TiO_2 is employed as the main photocatalytic agent; however, the addition of metals such as gold, silver, and copper to TiO_2 has also been considered, thus producing metal-induced photocatalysis, generating an important advantage in the development of more efficient composite materials in the face of visible light (Liu et al., 2017).

4.3 *Nanobioremediation, Interaction of Nanomaterials and Microorganisms*

The agrochemicals so necessary to ensure the development of crops and industrial development constitute one of the main causes of soil, water, and air pollution, considering the high volume of waste generated by them and that although in some countries, fortunately, the disposal and confinement procedures are well regulated, in the rest of the world there is a series of legal loopholes that are exploited, turning industrial waste into a source of pollution. Finally, another source of pollution is represented by medication used in the veterinary field. This is how three important areas for the social and economic development of the people become an ecological risk by releasing several organic compounds that end up compromising the safe consumption of water and wildlife associated with the various bodies of water. Physical and chemical methods have generated adequate results; however, the high cost they represent and their means of production can generate certain pollutants at the same time (Chaudhary et al., 2023). Therefore, one of the appropriate options is the use of microorganisms from the area that has pollution problems, trying to use this alternative decontamination friendly with the ecosystem, considering bioremediation as a suitable option.

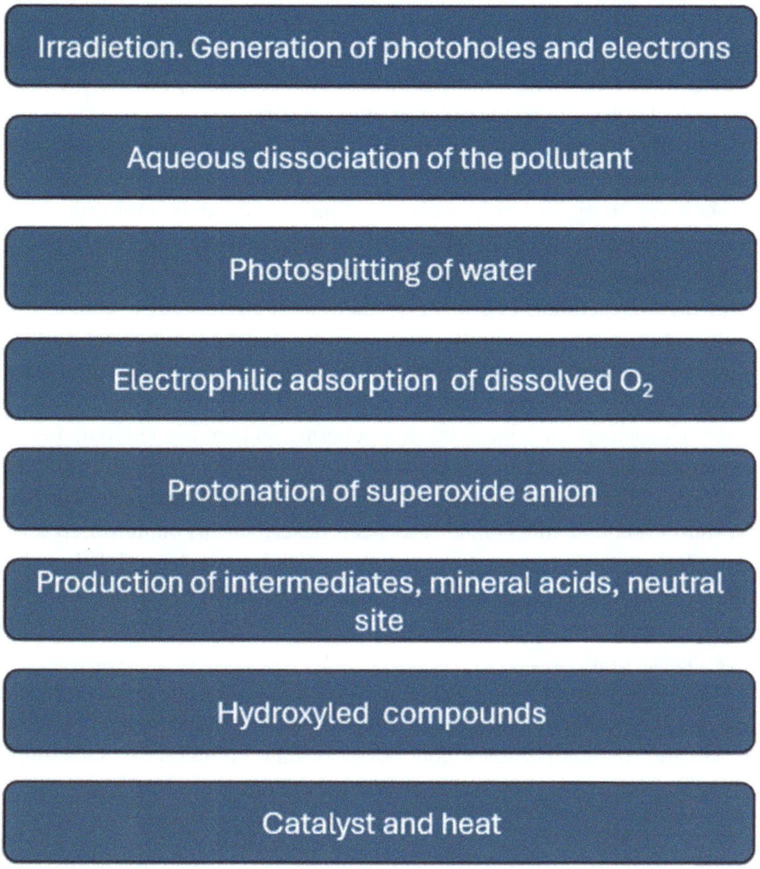

Fig. 3.2 Photocatalysis mechanism proposed by Pareek and Adesina

Although bioremediation is a suitable alternative to combat water and soil contamination, it is also necessary to consider the disadvantages of this ecological restoration procedure, such as a slow regeneration process, good results but in the long term, together with the fact that some halogenated contaminants and heavy-metal ions could considerably hinder the process (Kuppan et al., 2024). It is here that the application of nanomaterials can generate constructive interaction with microorganisms, enhancing the bioremediation process.

The coupling of nanoparticles with the biotic factors of a damaged ecosystem depends on several factors. First, the living being with which to generate the greatest interaction such as microorganisms (fungi and bacteria) and plants. The characteristics of the nanoparticle, such as its shape, size, surface coverage, and the chemical nature, constitute the second factor to consider. Other external factors also play an important role, such as the pH of the culture medium and the temperature. All these factors represent a series of combinations necessary to ensure a synergy between the microorganisms and the nanoparticles, thus enhancing the bioremediation effect. This is important to consider because the various interactions that would be generated could also be destructive, such as the fact that the nanoparticles could have antimicrobial activity, thus reducing the possibility of applying this type of technology (Vázquez-Núñez et al., 2020).

The role played by nanoparticles in interacting with microorganisms takes advantage of their characteristics such as surface area, their interaction with the surface, their reactivity, and the possibility of interacting at the molecular level with the microorganisms, facilitating the growth of the microorganism and the degradation of the contaminant.

One of the interaction mechanisms is that nanoparticles allow greater bioavailability of contaminants, especially when they are hydrophobic or are part of complex matrices, adsorbing these compounds and facilitating their access to microorganisms. On the other hand, the capacity of certain nanomaterials to generate reactive oxygen species, particles that generate an increase in the metabolism of microorganisms in response to oxidative stress, stimulating metabolic pathways that allow a more efficient degradation of contaminants, has also been reported (Akintola, 2024). Finally, another mechanism has been described, through which a positive interaction between microorganisms and nanoparticles is achieved. When the latter generate, by means of photocatalysis, a partial degradation of the pollutants occurs, thus increasing the bioavailability of simpler structures from the pollutants (Vázquez-Núñez et al., 2020).

The main nanomaterials used for nanobioremediation are nanocrystals, which facilitate enzymatic degradation by overexpressing organophosphate hydrolysates, Ti, Mn, Ag, and Au nanoparticles, nanopowders, nanomembranes, nanocomposites, nanotubes, and nanosponge in bacteria (Vázquez-Núñez et al., 2020).

5 Regulatory Frameworks and Ethical Considerations in Agricultural Nanotechnology

The rapid advancement of nanotechnology in agriculture has outpaced the development of comprehensive regulatory frameworks. Currently, there is a lack of standardized guidelines governing the use of nanomaterials in agricultural settings, leading to uncertainties regarding their safety and environmental impact. Some countries have initiated efforts to establish regulations, but a globally harmonized approach remains absent (Rai & Biswas, 2018).

Ethical considerations also emerge concerning the deployment of nanotechnology in agriculture. Issues such as informed consent, potential socioeconomic disparities, and the long-term sustainability of nanomaterial use necessitate transparent dialogue among stakeholders. Developing robust regulatory policies that address both safety and ethical concerns is crucial for the responsible integration of nanotechnology into agricultural practices (Ijaz et al., 2024).

6 Limitations of the Study

Despite the promising advances in nanotechnology-based sensors for detecting contaminants in agricultural soils and water, several limitations remain. First, the high cost and complexity of fabricating certain nanomaterials, such as metal = oxide nanoparticles and CNTs, can restrict their large-scale application, especially in resource-limited settings. Second, the long-term stability and durability of these sensors under diverse environmental conditions, including fluctuating temperatures, humidity, and soil composition, require further evaluation. Third, the potential toxicity and environmental impact of nanomaterials themselves need careful assessment to ensure safe use in agricultural ecosystems. Additionally, the integration of these sensors into existing agricultural practices demands user-friendly designs and robust data interpretation frameworks, which are still under development. Finally, many sensor platforms have been tested mainly under controlled laboratory conditions, and their effectiveness in complex, real-field environments remains to be thoroughly validated.

7 Conclusion

The contamination of agricultural soils by various pollutants poses a significant threat to crop productivity, food safety, and environmental health. Recent advances in nanotechnology have enabled the development of highly sensitive and selective sensors capable of detecting a wide range of contaminants, including heavy metals, pesticides, and pathogens, in soil environments. These nanotechnology-based sensors facilitate precision agriculture by enabling real-time monitoring, efficient resource management, and timely identification of harmful substances. This review discusses the current state of nanomaterial-enabled sensor technologies for soil contaminant detection, emphasizing their potential to improve agricultural productivity and sustainability through enhanced contaminant monitoring and management.

Nanotechnology-enabled sensors offer a transformative approach to the detection and management of contaminants in agricultural soils and water, enabling precision agriculture and sustainable resource use. These sensors demonstrate high sensitivity, selectivity, and rapid response in identifying heavy metals, pesticides, and microbial pathogens that threaten environmental and human health. The integration of such advanced sensing technologies can significantly enhance contaminant monitoring and timely intervention, contributing to improved crop productivity and food safety. However, challenges related to cost, environmental impact, and real-world applicability must be addressed through continued research and development. Overall, the deployment of nanomaterial-based sensors represents a promising strategy for advancing sustainable agriculture and protecting ecosystems from contamination.

References

Abdallat, G. A., Salameh, E., Shteiwi, M., & Bardaweel, S. (2022). Pharmaceuticals as emerging pollutants in the reclaimed wastewater used in irrigation and their effects on plants, soils, and groundwater. *Water (Switzerland), 14*(10). https://doi.org/10.3390/w14101560

Adam, T., & Dhahi, T. S. (2022). Nanosensors: Recent perspectives on attainments and future promise of downstream applications. *Process Biochemistry, 117*, 153–173. https://doi.org/10.1016/J.PROCBIO.2022.03.024

Akintola, A. A. (2024). Nanotechnology-enhanced microbial bioremediation of organic pollutants. *International Journal of Research Publication and Reviews Nanotechnology-Enhanced Microbial Bioremediation of Organic Pollutants*. https://doi.org/10.55248/gengpi.5.1024.2828

Alghuthaymi, M., Asran-Amal, M., et al. (2020). Carbon nanotubes: An efficient sorbent for herbicide sensing and remediation. In *Carbon nanomaterials for Agri-food and environmental applications* (pp. 429–457). https://doi.org/10.1016/B978-0-12-819786-8.00019-0

Alhalili, Z. (2023). Metal oxides nanoparticles: General structural description, chemical, physical, and biological synthesis methods, role in pesticides and heavy metal removal through wastewater treatment. *Molecules*. https://doi.org/10.3390/molecules28073086

Alshaybani, A. M., Bodowara, F. S., Al Mamlook, R., & Abdalrahmn, Z. M. (2024). Water disinfection via zinc oxide (ZnO) nanowires chemically fabricated on a modified polyurethane substrate. *International Journal of Environment, Engineering and Education, 6*(2), 73–83. https://doi.org/10.55151/ijeedu.v6i2.136

Antonio-Pérez, A., Durán-Armenta, L. F., Pérez-Loredo, M. G., & Torres-Huerta, A. L. (2023). Biosynthesis of copper nanoparticles with medicinal plants extracts: From extraction methods to applications. *Micromachines*. https://doi.org/10.3390/mi14101882

Arora, B., & Attri, P. (2020). Carbon nanotubes (CNTs): A potential nanomaterial for water purification. *Journal Of Composites Science*. https://doi.org/10.3390/jcs4030135

Asif, M., Yasmin, R., Asif, R., Ambreen, A., Mustafa, M., & Umbreen, S. (2022). Green synthesis of silver nanoparticles (AgNPs), structural characterization, and their antibacterial potential. *Dose-Response, 20*(1). https://doi.org/10.1177/15593258221088709

Baby, R., Hussein, M. Z., Abdullah, A. H., & Zainal, Z. (2022). Nanomaterials for the treatment of heavy metal contaminated water. *Polymers, 14*(3). https://doi.org/10.3390/polym14030583

Baruah, S., & Dutta, J. (2009). Nanotechnology applications in pollution sensing and degradation in agriculture. In *Environmental chemistry letters*. Springer Verlag. https://doi.org/10.1007/s10311-009-0228-8

Bounegru, A. V., Dinu Iacob, A., Iticescu, C., & Georgescu, P. L. (2025). Electrochemical sensors and biosensors for the detection of pharmaceutical contaminants in natural waters – A comprehensive review. *Chem*. https://doi.org/10.3390/chemosensors13020065

Castro-Muñoz, R., & Castro-Muñoz, R. (2020). The strategy of nanomaterials in polymeric membranes for water treatment: Nanocomposite membranes. *Tecnología y Ciencias Del Agua, 11*(1), 410–436. https://doi.org/10.24850/j-tyca-2020-01-11

Chaudhary, P., Ahamad, L., Chaudhary, A., Kumar, G., Chen, W. J., & Chen, S. (2023). Nanoparticle-mediated bioremediation as a powerful weapon in the removal of environmental pollutants. *Journal of Environmental Chemical Engineering, 11*(2), 109591. https://doi.org/10.1016/J.JECE.2023.109591

Choudhury, S. (2020). Molecular tools for the detection of waterborne pathogens. In *Waterborne pathogens: Detection and treatment* (pp. 219–235). https://doi.org/10.1016/B978-0-12-818783-8.00012-8

Dabas, S. (2025). Metal and metal oxide based nanoremediation: A sustainable alternative. *Discover Materials*. https://doi.org/10.1007/s43939-025-00242-6

Dafhalla, A. K. Y., Dhahi, T. S., Al-Mufti, A. W., Saad, S. A., Alqahtani, A. S., Al-nuaimi, M. A. H., Elobaid, M. E., Adam, T., & Gopinath, S. C. B. (2024). Nanogap nanowires and its applications in biosensing. *Sensing and Bio-Sensing Research, 44*, 100638.

Dager, A., Uchida, T., Maekawa, T., & Tachibana, M. (2019). Synthesis and characterization of mono-disperse carbon quantum dots from fennel seeds: Photoluminescence analysis using machine learning. *Scientific Reports, 9*(1). https://doi.org/10.1038/s41598-019-50397-5

Das, R., Abd Hamid, S. B., Ali, M. E., Ismail, A. F., Annuar, M. S. M., & Ramakrishna, S. (2014). Multifunctional carbon nanotubes in water treatment: The present, past and future. *Desalination, 354*, 160–179. https://doi.org/10.1016/J.DESAL.2014.09.032

de las Grijalva Endara, A., Jiménez Heinert, M. E., & Ponce Solórzano, H. X. (2020). Contaminación del agua y aire por agentes químicos. *RECIMUNDO, 4*(4), 79–93.

de Ramos Mancheno, A. D. (2024). Efectos del consumo de agua contaminada en la calidad de vida de las personas. *Polo Del Conocimiento, 9*(1), 614–632. https://doi.org/10.23857/pc.v9i1.6396

Devi, P., Rajput, P., Thakur, A., Kim, K. H., & Kumar, P. (2019). Recent advances in carbon quantum dot-based sensing of heavy metals in water. *TrAC – Trends in Analytical Chemistry, 114*, 171–195. https://doi.org/10.1016/j.trac.2019.03.003

Firozjaee, T. T., Nabi, R., Firozjaee, T. T., Mehrdadi, N., Baghdadi, M., & Nabi Bidhendi, G. R. (2018). Application of nanotechnology in pesticides removal from aqueous solutions-a review. *International Journal of Nanoscience and Nanotechnology, 14*.

Gelover Santiago, S. L. (2015). Nanotecnología, una alternativa para mejorar la calidad del agua. In I (Vol. 8, Issue 14).

Ghosh, N., Das, S., Biswas, G., & Haldar, P. K. (2022). Review on some metal oxide nanoparticles as effective adsorbent in wastewater treatment. In *Water science and technology*. IWA Publishing. https://doi.org/10.2166/wst.2022.153

Grozdanov, A., Dimitrievska, I., & Paunovic, P. (2023). Recent advancements in nano sensors for air and water pollution control. *Material Science & Engineering International Journal, 7*(2), 113–128.

Hammami, I., Alabdallah, N. M., Jomaa, A., et al. (2021). Gold nanoparticles: Synthesis properties and applications. *Journal of King Saud University - Science, 33*(7). https://doi.org/10.1016/j.jksus.2021.101560

He, Z., Yin, H., Chang, C. C., Wang, G., & Liang, X. (2020). Interfacing DNA with gold nanoparticles for heavy metal detection. *Biosensors*. https://doi.org/10.3390/BIOS10110167

Hemlata, M., et al. (2020). Biosynthesis of silver nanoparticles using Cucumis prophetarum aqueous leaf extract and their antibacterial and Antiproliferative activity against cancer cell lines. *ACS Omega, 5*(10), 5520–5528. https://doi.org/10.1021/acsomega.0c00155

Hussain, S., Naeem, M., & Chaudhry, M. N. (2017). Estimation of residual antibiotics in soil and underground water of areas affected by pharmaceutical wastewater in Lahore. *Journal of Water Chemistry and Technology, 39*(1), 56–60. https://doi.org/10.3103/S1063455X1701009X

Ijaz, S., Iqbal, J., Abbasi, B. A., Ullah, Z., Ijaz, N., Yaseen, T., et al. (2024). Regulatory and ethical concerns of nanotechnology. In R. Shahzad, S. Fiaz, et al. (Eds.), *Agriculture BT – Revolutionizing agriculture: A comprehensive exploration of Agri-nanotechnology* (pp. 395–427). Springer Nature Switzerland. https://doi.org/10.1007/978-3-031-76000-6_18

Javed, R., Zia, M., Naz, S., Aisida, S. O., Ain, N., et al. (2020). Role of capping agents in the application of nanoparticles in biomedicine and environmental remediation: Recent trends and future prospects. *Journal of Nanobiotechnology, 18*(1), 1–15. https://doi.org/10.1186/s12951-020-00704-4

Kanoun, O., Lazarević-Pašti, T., Pašti, I., Nasraoui, S., Talbi, M., Brahem, A., Adiraju, A., Sheremet, E., Rodriguez, R. D., Ali, M. B., & Al-Hamry, A. (2021). A review of nanocomposite-modified electrochemical sensors for water quality monitoring. *Sensors, 21*(12). https://doi.org/10.3390/s21124131

Kausar, A., & Ahmad, I. (2024). Highpoints of carbon nanotube nanocomposite sensors – A review. *E-Prime – Advances in Electrical Engineering, Electronics and Energy, 7*, 100419. https://doi.org/10.1016/J.PRIME.2024.100419

Kayode-Afolayan, S. D., Ahuekwe, E. F., & Nwinyi, O. C. (2022). Impacts of pharmaceutical effluents on aquatic ecosystems. *Scientific African*. https://doi.org/10.1016/j.sciaf.2022.e01288

Kedar, K., Nayak, S., Bhaskar, V., & Dist Thane, M. (2022). *Synthesis of Silver Nanoparticles by Chemical Reduction Method INTRODUCTION* (Vol. 25, Issue 3). www.ijppr.humanjournals.comwww.ijppr.humanjournals.com

Khraisheh, M., Elhenawy, S., Almomani, F., Al-Ghouti, M., Hassan, M. K., & Hameed, B. H. (2021). Recent progress on nanomaterial-based membranes for water treatment. *Membranes*. https://doi.org/10.3390/membranes11120995

Kolluru, S. S., Agarwal, S., Sireesha, S., Sreedhar, I., & Kale, S. R. (2021). Heavy metal removal from wastewater using nanomaterials-process and engineering aspects. *Process Safety and Environmental Protection, 150*, 323–355. https://doi.org/10.1016/J.PSEP.2021.04.025

Kuppan, N., Padman, M., Mahadeva, M., Srinivasan, S., & Devarajan, R. (2024). A comprehensive review of sustainable bioremediation techniques: Eco friendly solutions for waste and pollution management. *Waste Management Bulletin, 2*(3), 154–171. https://doi.org/10.1016/j.wmb.2024.07.005

Lariza, M., Cuel, A., Itzel, L., López, L., & Calindo, A. S. (2012). *Nanotubos de carbono: funcionalización y aplicaciones biológicas Carbon nanotubes: Funcionalization and biological applications.*

Liu, L., Zhang, X., Yang, L., Ren, L., Wang, D., & Ye, J. (2017). Metal nanoparticles induced photocatalysis. *National Science Review, 4*(5), 761–780. https://doi.org/10.1093/nsr/nwx019

Liu, Y., Xue, Q., Chang, C., Wang, R., Liu, Z., & He, L. (2022). Recent progress regarding electrochemical sensors for the detection of typical pollutants in water environments. *Analytical Sciences, 38*(1), 55–70. https://doi.org/10.2116/analsci.21SAR12

Long, C., Jiang, Z., Shangguan, J., Qing, T., Zhang, P., & Feng, B. (2021). Applications of carbon dots in environmental pollution control: A review. *Chemical Engineering Journal, 406*. https://doi.org/10.1016/j.cej.2020.126848

Markina, N. E., Ustinov, S. N., Zakharevich, A. M., & Markin, A. V. (2020). Copper nanoparticles for SERS-based determination of some cephalosporin antibiotics in spiked human urine. *Analytica Chimica Acta, 1138*, 9–17. https://doi.org/10.1016/J.ACA.2020.09.016

Maubert, M., Soto, L., León, A. M., & Flores, J. (2009). Nanotubos de Carbono-La era de la nanotecnología. *Razón y palabra, 68*.

McGarrity, M., & Zhao, F. (2023). Graphene-based Chemiresistor sensors for drinking water quality monitoring. *Sensors, 23*(24). https://doi.org/10.3390/s23249828

Naseem, T., & Durrani, T. (2021, January 1). The role of some important metal oxide nanoparticles for wastewater and antibacterial applications: A review. *Environmental Chemistry and Ecotoxicology*. https://doi.org/10.1016/j.enceco.2020.12.001

Noriega-Treviño, M., Quintero González, E., Cristina, C., Pacheco, G. et al (2012). *Tecnología y Ciencias del Agua*. http://www.redalyc.org/articulo.oa?id=353531980006

Nzilu, D. M., Madivoli, E. S., Makhanu, D. S., Wanakai, S. I., Kiprono, G. K., & Kareru, P. G. (2023). Green synthesis of copper oxide nanoparticles and its efficiency in degradation of rifampicin antibiotic. *Scientific Reports, 13*(1). https://doi.org/10.1038/s41598-023-41119-z

Pabón, S. E., Benítez, R., Sarria-Villa, R. A., & Gallo, J. A. (2020). Contaminación del agua por metales pesados, métodos de análisis y tecnologías de remoción. *Entre Ciencia e Ingeniería, 14*, 9–18.

Rai, M., & Biswas, J. K. (2018). *Nanomaterials: Ecotoxicity, safety, and public perception.*

Ríos-Tobón, S., Agudelo-Cadavid, R. M., & Gutiérrez-Builes, L. A. (2017). Pathogens and microbiological indicators of the quality of water for human consumption. *Revista Facultad Nacional de Salud Publica, 35*(2), 236–247. https://doi.org/10.17533/udea.rfnsp.v35n2a08

Saldivar, L., & Walsh, C. (2015). Nanotecnología para el tratamiento de agua. Claves sobre la investigación en México 1. In www.mundonano.unam.mx | (Vol. 8, Issue 14). http://www.scidev.net/en/new-technologies/nanotechnology-for-*SDG-6-Summary-Progress-Update-2021_Version-July-2021_SP* (n.d.).

Sener, G. (2014). Colorimetric sensor array based on gold nanoparticles and amino acids for identification of toxic metal ions in water. *ACS Applied Materials and Interfaces, 6*(21), 18395–18400. https://doi.org/10.1021/am5071283

Shantkriti, S., & Rani, P. (2014). Biological synthesis of copper nanoparticles using Pseudomonas fluorescens. *International Journal of Current Microbiology and Applied Sciences, 3*(9), 374–383.

Shin, H., Lee, J. S., Kim, D. H., Seo, J. W., Choi, S. J., & Kim, I. D. (2025). Chemiresistors. In *Nature-inspired sensors* (pp. 295–319). https://doi.org/10.1016/B978-0-443-15684-7.00026-9

Sugunan, A., Thanachayanont, C., Dutta, J., & Hilborn, J. G. (2005). Heavy-metal ion sensors using chitosan-capped gold nanoparticles. *Science and Technology of Advanced Materials, 6*(3–4), 335–340. https://doi.org/10.1016/j.stam.2005.03.007

Tejaswi, T. S., & Devi, P. S. (2025). Biogenesis of copper oxide nanoparticles using marine bacteria Pseudomonas aeruginosa: In vitro antimicrobial and photocatalytic activities. *Journal of pure and applied. Microbiology, 19*(1), 210–226. https://doi.org/10.22207/JPAM.19.1.12

Teng, Y., Zhong, Y., Xu, P., Li, J., Pan, Z., Hu, T., et al. (2025). Utilizing pillararenes as capping agents to stabilize copper nanoparticles for cost-effective and high-performance SERS application. *Spectrochimica Acta Part A: Molecular and Biomolecular Spectroscopy, 334*, 125919. https://doi.org/10.1016/J.SAA.2025.125919

Vaseashta, A., Vaclavikova, M., Vaseashta, S., Gallios, G., Roy, P., & Pummakarnchana, O. (2007). Nanostructures in environmental pollution detection, monitoring, and remediation. *Science and Technology of Advanced Materials, 8*(1–2), 47–59. https://doi.org/10.1016/j.stam.2006.11.003

Vazquez-Duhalt, R., & Vazquez-Duhalt, R. (2015). Nanotecnología en procesos ambientales y remediación de la contaminación. *Mundo Nano. Revista Interdisciplinaria En Nanociencias y Nanotecnología, 8*(14), 70–80. https://doi.org/10.22201/ceiich.24485691e.2015.14.52514

Vázquez-Núñez, E., Molina-Guerrero, C. E., Peña-Castro, J. M., Fernández-Luqueño, F., & de la Rosa-Álvarez, M. G. (2020). Use of nanotechnology for the bioremediation of contaminants: A review. *PRO, 8*(7), 1–17. https://doi.org/10.3390/pr8070826

Vikesland, P. J. (2018). Nanosensors for water quality monitoring. *Nature Nanotechnology, 13*(8), 651–660. https://doi.org/10.1038/s41565-018-0209-9

Wang, C., & Yu, C. (2013). Detection of chemical pollutants in water using gold nanoparticles as sensors: A review. *Reviews in Analytical Chemistry, 32*(1), 1–14. https://doi.org/10.1515/revac-2012-0023

Xu, X., Ray, R., Gu, Y., Ploehn, H. J., Gearheart, L., Raker, K., & Scrivens, W. A. (2004). Electrophoretic analysis and purification of fluorescent single-walled carbon nanotube fragments. *Journal of the American Chemical Society, 126*(40), 12736–12737. https://doi.org/10.1021/ja040082h

Yan, F., Sun, Z., Zhang, H., Sun, X., Jiang, Y., & Bai, Z. (2019). The fluorescence mechanism of carbon dots, and methods for tuning their emission color: A review. *Microchimica Acta, 186*(8). https://doi.org/10.1007/s00604-019-3688-y

Yunus, I. S., Harwin, K., et al. (2012). Nanotechnologies in water and air pollution treatment. *Environmental Technology Reviews, 1*(1), 136–148. https://doi.org/10.1080/21622515.2012.733966

Zaporotskova, I. V., Boroznina, N. P., Parkhomenko, Y. N., & Kozhitov, L. V. (2016). Carbon nanotubes: Sensor properties. A review. *Modern Electronic Materials, 2*(4), 95–105. https://doi.org/10.1016/j.moem.2017.02.002

Zhu, S., Song, Y., Zhao, X., Shao, J., Zhang, J., & Yang, B. (2015). The photoluminescence mechanism in carbon dots (graphene quantum dots, carbon nanodots, and polymer dots): Current state and future perspective. *Nano Research, 8*(2), 355–381. https://doi.org/10.1007/s12274-014-0644-3

Zubiarrain-Laserna, A., & Kruse, P. (2020). Review – Graphene-based water quality sensors. *Journal of the Electrochemical Society, 167*(3), 037539. https://doi.org/10.1149/1945-7111/ab67a5

Zwolak, A., Sarzyńska, M., Szpyrka, E., & Stawarczyk, K. (2019). Sources of soil pollution by heavy metals and their accumulation in vegetables: A review. *Water, Air, and Soil Pollution, 230*(7). https://doi.org/10.1007/s11270-019-4221-y

Chapter 4
Autologous Peripheral Blood Products in Regenerative Medicine: Preparation, Mechanisms, and Applications

Paulina Delgado-González, David A. de la Garza-Kalife, Adriana G. Quiroz-Reyes, and Elsa N. Garza-Treviño

1 Introduction

Autologous peripheral blood products (APBPs) have emerged as promising, natural alternatives to conventional pharmacological treatments for several medical conditions. These therapies, often called autohemotherapy, involve using components derived from a patient's blood to stimulate healing and modulate disease processes. Historically applied in dermatology during the early twentieth century, autohemotherapy fell out of favor owing to limited scientific validation (Brewer & Brewer, 2014). However, renewed clinical interest, driven by advances in regenerative medicine and the growing demand for personalized therapies, has sparked a resurgence in research and application.

This interest aligns with broader shifts toward minimally invasive, patient-specific treatments. In orthopedics and sports medicine, for example, APBPs such as platelet-rich plasma (PRP) and autologous conditioned serum (ACS) are being used to treat osteoarthritis (OA) and musculoskeletal injuries, with encouraging outcomes (Gupta & Jain, 2024b). These blood-based products harness endogenous growth factors and cytokines to promote regeneration, reduce inflammation, and improve pain and function (Mayoral Rojals et al., 2023). OA, the most prevalent joint disease globally, exemplifies a condition where such innovative strategies are in demand—especially in cases where standard therapies are ineffective (Quicke et al., 2022).

P. Delgado-González · D. A. de la Garza-Kalife · A. G. Quiroz-Reyes ·
E. N. Garza-Treviño (✉)
Faculty of Medicine, Department of Biochemistry and Molecular Medicine, Cell Therapy Laboratory, Autonomous University of Nuevo Leon (UANL), Monterrey, Mexico
e-mail: paulina.delgadogn@uanl.edu.mx; david.delagarzaka@uanl.edu.mx; guadalupe.quirozrys@uanl.edu.mx; elsa.garzatr@uanl.edu.mx

M. S. Heya (ed.), *Natural Alternatives and Nanotechnology Applied to the One Health Approach*, https://doi.org/10.1007/978-3-032-03675-9_4

In elite sports, PRP is increasingly favored for treating conditions such as patellar tendinopathy and early OA because of its high concentration of bioactive molecules that promote wound healing, neovascularization, and tissue remodeling, making it a focal point of regenerative therapies (Foster et al., 2009; Pretorius et al., 2023). A Delphi consensus among NFL physicians from 26 teams supported its use for these indications, highlighting the need for caution owning to the proliferation of unregulated biologics lacking standardization or evidence (Murray et al., 2023). This underscores a broader challenge: ensuring therapeutic rigor, safety, and reproducibility in autologous products.

The objective of this chapter is to explore the mechanisms, current clinical uses, and future directions of autologous peripheral blood-derived therapies. Special attention will be given to innovations such as natural compounds that may enhance therapeutic efficacy and their potential role in One Health, expanding their utility across diverse medical contexts.

2 Autologous Peripheral Blood Products

Autologous peripheral blood products have demonstrated significant therapeutic benefits in musculoskeletal injuries, chronic wounds, and degenerative joint diseases, largely because of the bioactive molecules naturally present in the blood, such as growth factors and cytokines. Beyond these, components such as antimicrobial peptides (AMPs) and neutrophil degranulation products also play crucial roles in modulating immune responses and promoting wound healing. Various APBPs are employed in regenerative medicine, each prepared through specific protocols that optimize their composition and therapeutic potential (see Table 4.1). Additionally, Fig. 4.1 illustrates the diverse preparation methods, emphasizing differences in product derivation, centrifugation speeds, and activation techniques. To understand how autologous blood therapies exert their effects, it is important to first consider the fundamental components of blood as a tissue.

2.1 Blood Components

Blood is a fluid connective tissue composed of plasma and cellular elements, with plasma accounting for roughly 55% of its volume. Primarily water, plasma is a transport medium for proteins, electrolytes, gases, nutrients, and metabolic waste. Its major proteins include albumin, globulins, and fibrinogen, and when clotting factors are removed, the remaining fluid is known as serum. The remaining 45% of blood volume comprises cellular components: erythrocytes (red blood cells), leukocytes (white blood cells), and platelets. Each cell type plays a distinct role in health, repair, and immune regulation, which supports the rationale behind APBPs. Erythrocytes transport oxygen and modulate oxidative stress, leukocytes mediate

Table 4.1 Overview of autologous peripheral blood products, their preparation protocols, and reasoning behind their proposed composition

Product	Description	Preparation	Components	Advantages
PRP	APBP derived from the plasma of centrifuged blood	Blood in the ACD tube is centrifuged at 3200 rpm for 4 min—the lower third of plasma is PRP	Rich in growth factors such as PDGF, VEGF, TGF-β1, IGF-1, EGF, and FGF	Higher platelet concentration than blood levels, which may be leukocyte-poor or -rich, with various medical applications
PRGF	PRGF is a platelet-rich APBP with lower platelet levels than PRP and no leukocytes	Citrate-anticoagulated blood is centrifuged at 1850 rpm for 8 min, dividing into three plasma layers, and then activated with $CaCl_2$	Eliminates leukocytes to reduce pro-inflammatory activity, but ironically, contains lower levels of growth factors (TGF-β1, PDGF, VEGF) and pro-inflammatory cytokines (IL-1β) than other APBPs	Simplified preparation improves reproducibility by using $CaCl_2$ instead of thrombin to activate clotting
PL	PL is made by freeze-thawing PRP to rupture platelets and release growth factors from their α-granules	3–5 cycles of freezing PRP at −80 °C for 30 min, thawing at 37 °C, and then centrifuging at 4000 rpm for 10 min	Like PRP, PL contains IGF-1, PDGF isoforms (AA, BB, AB), TGF-β1, BDNF, EGF, VEGF, HGF, and FGF—with some studies reporting elevated TGF-β1, PDGF-AB, IGF-1, and BDNF	PL enables cryopreservation and long-term storage, overcoming the limitations of PRP related to donor variability, inconsistent growth factor content, and incomplete platelet degranulation
PRF	Fibrin matrix that releases platelet-derived growth factors over a longer duration than PRP	Immediately centrifuged blood at 2700 rpm for 10 min forms a PRF clot, whereas lower centrifugation speeds form softer fibrin clots termed A-PRF and i-PRF	A leukocyte-rich fibrin matrix forms by platelet activation and fibrin polymerization, enabling the sustained release of PDGF, TGF-β1, VEGF, BMP-2, EGF, and IGF-1	Although PRF is a platelet concentrate, its primary effect stems from its slowly polymerized fibrin matrix, which forms a scaffold that supports tissue healing
CGFs	Modified PRF product obtained through alternating centrifugation speeds resulting in a rigid fibrin matrix rich in growth factors	Blood is processed by programmed centrifugation of 2 min at 2700 rpm, 4 min at 2400 rpm, 4 min at 2700 rpm, and 3 min at 3000 rpm, yielding a central buffy coat in a middle CGF gel layer	Fibrinolytic activation of platelets embedded in the CGF fibrin scaffold enables the sustained release of key growth factors—TGF-β1, PDGF-BB, IGF-1, BMP, VEGF, EGF, and FGF	Like PRF, CGF stimulates a sustained release of growth factors over 14 days; however, it demonstrates superior tensile strength, higher growth factor content, and greater long-term bone induction and tissue regeneration

(continued)

Table 4.1 (continued)

Product	Description	Preparation	Components	Advantages
HS	HS, also known as serum from PRF, is a cell- and fibrin-free blood derivative obtained by squeezing serum from the PRF clot	Preparation protocol of HS is similar to that of PRF, followed by squeezing of the serum from the formed PRF clot at the end of the procedure	HS-derived EVs modulate IL-6 and TGF-β/SMAD signaling, promoting tissue remodeling and immune regulation, also containing antiangiogenic factors PF-4, serpin, and TIMP-1	A liquid without coagulation factors that gives the final products a better cell proliferative effect, showing promising results in enhancing cartilage cell viability and reducing inflammation in OA
ACS	Blood exposed to $CrSO_4$-coated glass beads, resulting in serum enriched with anti-inflammatory cytokines	Blood is incubated with $CrSO_4$-coated glass beads and no anticoagulant at 37 °C for 24 h, followed by centrifugation at 3200 rpm for 10 min	ACS outperforms PRP in pain relief, attributed to higher concentrations of PDGF-BB and anti-inflammatory cytokines such as IL-1Ra, IL-4, IL-10, and IL-13	ACS offers an advantage over PRP by specifically targeting inflammatory pathways through its enriched anti-inflammatory cytokines
APS	L-PRP exposed to polyacrylamide beads, resulting in plasma enriched with anti-inflammatory cytokines	ACD-treated blood is centrifuged at 2000 rpm for 15 min to isolate L-PRP, which is exposed to polyacrylamide beads and centrifuged again for 2 min	Contact between the blood and polyacrylamide beads stimulates the production of numerous cytokines and growth factors including sTNF-R1, TGF-β, IL-1Ra, PDGF-AB/BB, IGF-1, and EGF	APS was developed to target inflammatory pathways mediated by IL-1β and TNF-α, proposing to increase its clinical use with a higher concentration of leukocytes and platelets

ACD acid citrate dextrose, *ACS* autologous conditioned serum, *APBP* autologous peripheral blood products, *APS* autologous protein solution, *A-PRF* advanced form of platelet-rich fibrin, *CGF* concentrated growth factor, *HS* hyperacute serum, *i-PRF* injectable platelet-rich fibrin, *L-PRP* leukocyte-rich platelet-rich plasm, *OA* osteoarthritis, *PL* platelet lysate, *PRF* platelet-rich fibrin, *PRGF* plasma rich in growth factors, *PRP* platelet-rich plasma

immune defense, and platelets are central to hemostasis. Collectively, these cells sustain vital functions essential to human life, including oxygen delivery, immune protection, blood clotting, and tissue maintenance (Kuhn et al., 2017).

Erythrocytes, the most abundant blood cells, make up approximately 94% of the cellular volume of blood. These anucleate, organelle-free cells rely on glycolysis for energy throughout their 120-day lifespan and possess a structurally resilient, flexible biconcave shape that maximizes surface area for efficient gas exchange (Zhang et al., 2024). Erythrocytes are densely packed with hemoglobin, a tetrameric protein that enables them to transport oxygen from the lungs to tissues and carry carbon dioxide back for exhalation, with oxygen affinity dynamically regulated by pH, temperature, and 2,3-biphosphoglycerate (2,3-BPG) through the Bohr and Haldane effects (Glenn & Armstrong, 2019).

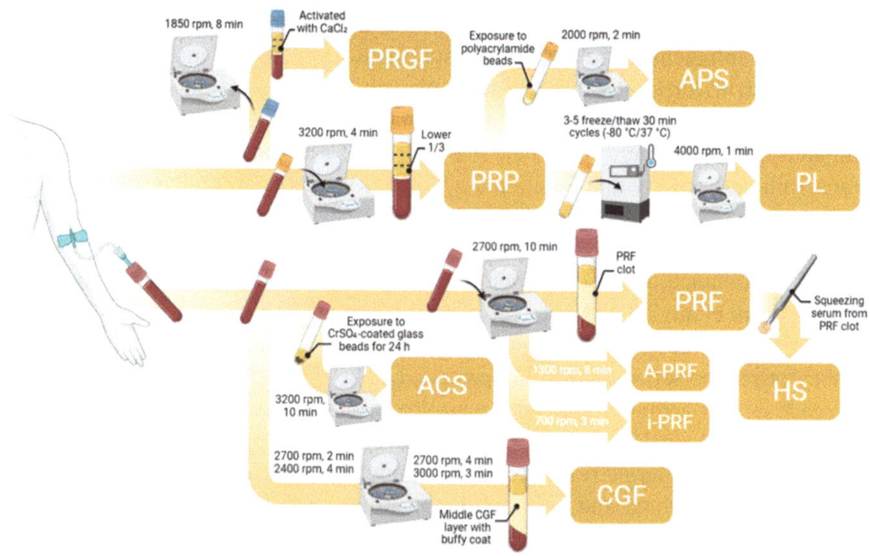

Fig. 4.1 Overview of diverse preparation methods for autologous peripheral blood products. Summary of various preparation techniques, emphasizing differences in tube selection (with or without anticoagulants), centrifugation protocols (speed and duration), and downstream processing steps such as activation or freezing. Arrows indicate the progression from whole blood to final products, illustrating shared steps and points of divergence in preparation. *ACS* autologous conditioned serum, *APS* autologous protein solution, *A-PRF* advanced form of platelet-rich fibrin, *CGF* concentrated growth factor, *HS* hyperacute serum, *i-PRF* injectable platelet-rich fibrin, *PL* platelet lysate, *PRF* platelet-rich fibrin, *PRGF* plasma rich in growth factors, *PRP* platelet-rich plasma

Erythropoiesis takes place in the bone marrow and is primarily regulated by erythropoietin, a hormone produced by the kidneys in response to hypoxemia. Hematopoietic stem cells differentiate into erythroblasts, which progressively shrink, synthesize hemoglobin, and expel their nuclei to become reticulocytes—immature red blood cells that complete maturation in the bloodstream within 24–48 h (Minetti et al., 2021; Moras et al., 2017).

Leukocytes constitute about 1% of blood volume but play a vital role in immune defense. In adults, leukocytes originate from hematopoietic stem cells in the bone marrow, which differentiate into myeloid or lymphoid lineages under the influence of cytokines and growth factors (Cabezas-Wallscheid et al., 2014; Pietras et al., 2015). Based on their lineage and granule content, leukocytes are categorized into granulocytes (neutrophils, eosinophils, basophils) and agranulocytes (lymphocytes, monocytes), with each subtype performing specialized functions shaped by its structure, lifespan, and responsiveness to distinct immune challenges (Summers et al., 2010).

Neutrophils, the most abundant leukocytes, are rapid responders in acute inflammation and microbial defense. Characterized by multilobed nuclei, neutrophils eliminate pathogens at infection sites through degranulation and phagocytosis. Their primary (azurophilic) granules contain myeloperoxidase, acid hydrolases, defensins, and AMPs such as cathelicidin; secondary granules house enzymes such as type IV collagenase, gelatinase, phospholipase, and AMPs including lysozyme and lactoferrin; tertiary granules contain phosphatases and metalloproteinases. The diversity of granules reflects the multifunctional roles of neutrophils in innate immunity. Despite their short half-life (6–12 h), neutrophils display remarkable transcriptional activity and functional adaptability, engaging in phagocytosis, reactive oxygen species production, cytokine secretion, and crosstalk with other immune cells. Recent studies suggest their involvement in innate immune memory and tumor surveillance, expanding their known immunological roles (Bowers et al., 2018; Martin et al., 2003; Noetzli et al., 2019).

Eosinophils, recognizable by their bilobed nuclei and eosinophilic granules, contain cytotoxic proteins major basic protein, eosinophil cationic protein, eosinophil peroxidase, and eosinophil-derived neurotoxin that target protozoa and helminths (Nussbaum et al., 2013; Valent et al., 2012). In addition to parasitic defense, they are key mediators of allergic responses and chronic inflammation, contributing to conditions such as asthma and eosinophilic gastrointestinal disorders. Their granules also contain enzymes such as histaminase and arylsulfatase, which help to regulate the tissue-damaging effects of other inflammatory mediators (Klion et al., 2020).

Basophils, the least abundant granulocytes (<0.5% of leukocytes), play a central role in hypersensitivity reactions and chronic inflammation. Their large basophilic granules release histamine and heparan sulfate, which increase vascular permeability and recruit immune cells; heparin, an anticoagulant; and cytokines such as interleukin 4 (IL-4) and interleukin 13 (IL-13), which drive immunoglobulin E (IgE) synthesis. Beyond their effector functions, basophils modulate T-helper cell responses and contribute to IgE- and immunoglobulin G (IgG)-mediated anaphylaxis (Dwyer et al., 2016; Shah et al., 2021).

Monocytes, characterized by their indented nuclei, circulate in the bloodstream before migrating into tissues, in which they differentiate into macrophages or other mononuclear phagocyte system cells. These macrophages perform key roles in phagocytosis, antigen presentation, and cytokine secretion (Coillard & Segura, 2019; Dash et al., 2024). Tissue-resident macrophages such as osteoclasts in bones and Kupffer cells in the liver (Hoeffel & Ginhoux, 2018; Italiani & Boraschi, 2015; Wu & Hirschi, 2021) adopt specialized functions and can polarize into pro-inflammatory (M1) or pro-resolving (M2) phenotypes in response to local signals, influencing processes such as fibrosis, immune regulation, and wound healing (Italiani & Boraschi, 2014).

Lymphocytes are central effectors of the immune system, distinguished by their dense, spheroidal nuclei and minimal cytoplasm. They include B cells, T cells, and natural killer (NK) cells, each playing distinct roles in immune defense. B cells mature in the bone marrow and mediate humoral immunity by producing antibodies; T cells mature in the thymus and contribute to cell-mediated immunity through helper and cytotoxic subsets; NK cells, part of the innate immune system, provide rapid cytotoxic responses against virus-infected and malignant cells without prior antigen exposure (Mace, 2023).

Platelets are small (2–3 μm), anucleate cytoplasmic fragments that circulate in an inactive state and rapidly activate in response to vascular injury. Derived from large, polyploid megakaryocytes of myeloid origin in the bone marrow, platelets have a lifespan of 8–10 days in the bloodstream. They are structurally organized into four functional zones:

1. The peripheral zone, consisting of the plasma membrane and glycocalyx enriched with glycoproteins, glycosaminoglycans, and coagulation factors; the sol-gel zone, containing microtubules, actin filaments, myosin, and structural proteins;
2. The organelle zone, which houses mitochondria, peroxisomes, glycogen, and three types of granules: α-granules (fibrinogen, coagulation factors, plasminogen, and platelet-derived growth factor [PDGF]), δ-granules (adenosine diphosphate [ADP], adenosine triphosphate [ATP], serotonin, and histamine), and λ-granules (hydrolytic enzymes);
3. The membranous zone, composed of the open canalicular and dense tubular systems involved in calcium storage and release.

Upon activation, platelets adhere to exposed collagen and von Willebrand factor, release granule contents (e.g., serotonin, ADP, thromboxane A_2), aggregate via glycoprotein IIb/IIIa–fibrinogen binding, and externalize phosphatidylserine to promote coagulation (Yun et al., 2016). Beyond hemostasis, platelets modulate immune responses through interactions with leukocytes and secretion of pro-inflammatory mediators, implicating them in host defense and making antiplatelet strategies relevant in conditions such as sepsis (Thomas & Storey, 2015).

Understanding blood composition lays the groundwork for appreciating how APBPs harness these elements for therapeutic purposes. By isolating and concentrating specific components—such as platelets in PRP or cytokines in conditioned serum—clinicians can amplify the body's natural repair mechanisms. These therapies exemplify a targeted use of the body's ability to address injury and disease, offering a personalized, minimally invasive alternative to conventional treatments. The following sections provide an overview of key APBPs, each with distinct compositions, preparation methods, and clinical applications—that contribute to the expanding landscape of regenerative and anti-inflammatory therapies.

2.2 Platelet-Rich Plasma

Platelet-rich plasma is an APBP derived from the plasma fraction of whole blood following centrifugation, characterized by a platelet concentration higher than normal physiological levels (Collins et al., 2021). However, its therapeutic efficacy and standardization remain complicated by variations in preparation, leukocyte content, and activation status (Ragni et al., 2025). The first PRP preparation methods were introduced in the 1970s for use in thrombocytopenic patients. Since then, numerous variations have been developed and adapted across diverse medical fields, with a notable surge in musculoskeletal applications (Collins et al., 2021). These protocols differ based on factors such as centrifugation speed and duration, temperature settings, choice of anticoagulants, and the volume of blood processed.

Standard PRP preparation involves collecting whole blood in anticoagulant-containing tubes and using differential centrifugation—a process that separates blood components into three layers: an upper layer that contains mostly platelets and leukocytes, an intermediate thin layer that is known as the buffy coat and that is rich in leukocytes, and a bottom layer that consists mostly of erythrocytes (Davis et al., 2014). Dhurat & Sukesh describe a PRP preparation method in which whole blood is collected in acid citrate dextrose (ACD) tubes and undergoes two centrifugation steps at 3200 rpm, 1800 *g* for 4 min: the first separates the erythrocytes, and the second, at a higher speed, concentrates platelets. The resulting plasma is divided, with the lower third containing PRP and the upper two-thirds forming platelet-poor plasma (PPP), which is discarded to isolate the PRP (Dhurat & Sukesh, 2014).

According to DeLong et al., PRP preparations can be classified into plasma-derived or buffy-coat-derived types based on their origin, with the key distinction being leukocyte content, leading to further categorization into leukocyte-poor PRP (P-PRP) and leukocyte-rich PRP (L-PRP) (Delong et al., 2012; Dohan Ehrenfest et al., 2014). P-PRP, such as autologous conditioned plasma, is typically prepared using a single-spin centrifugation method that collects the upper plasma layer while minimizing leukocyte inclusion, following standardized manufacturer protocols (Gupta & Jain, 2024a). In contrast, to produce L-PRP, the entire buffy coat and some erythrocytes are included (Dhurat & Sukesh, 2014). Many PRP preparation protocols add an activation step before administration, typically involving the addition of thrombin and/or calcium chloride ($CaCl_2$). Regardless, some clinicians administer PRP without pre-activation, instead relying on the natural activation of platelets triggered by their contact with collagen in the body's connective tissues (Di Matteo et al., 2015).

Blood-derived products exert therapeutic effects through a complex interaction of bioactive compounds that influence cellular behavior. The elevated concentration of platelets, which release various growth factors, cytokines, and extracellular vesicles (EVs), present in PRP, plays a critical role in mediating the therapeutic effects observed in musculoskeletal injuries, skin-wound healing, and tissue regeneration (Do Amaral et al., 2016; Patel et al., 2023). Rich in growth factors such as PDGF, vascular endothelial growth factor (VEGF), transforming growth factor-beta 1 (TGF-β1), insulin-like growth factor 1 (IGF-1), epidermal growth factor (EGF), and fibroblast growth factor (FGF), PRP promotes cell proliferation, angiogenesis, and tissue repair (Gharpinde et al., 2024; Oneto et al., 2020). Studies have shown that these factors can accelerate healing by promoting the migration of fibroblasts and endothelial cells into the injured tissue (Bendinelli et al., 2010). The growth factors in PRP enhance tenocyte proliferation, collagen synthesis, and tendon matrix regeneration (Kale et al., 2024). Although the precise mechanisms remain unclear, PRP appears to exert regenerative and anti-aging effects on the skin by releasing growth factors that stimulate DNA synthesis, gene expression, collagen and elastin deposition, and mitigate UV-induced damage through modulation of matrix metalloproteinase 1 (MMP-1), tyrosinase, and extracellular matrix proteins (Manole et al., 2024).

Research has illustrated that PRP-derived EVs can actively participate in wound healing, promoting re-epithelialization and modulating the immune response, a natural process in chronic wound management (Guo et al., 2017; Wu et al., 2021). These vesicles carry proteins and RNA that amplify the roles of growth factors and can exert paracrine effects on neighboring cells, thus further facilitating the repair mechanisms (Guo et al., 2017; Wu et al., 2021). In models of glucocorticoid-induced osteonecrosis of the femoral head, PRP-derived exosomes have been shown to counteract stress-mediated apoptosis by activating the Akt/Bad/Bcl-2 pathway and promoting angiogenesis (Tao et al., 2017). Platelet-derived EVs, naturally in circulation, are particularly effective in regenerative therapies owing to their low immunogenicity, tumorigenicity, and ability to modulate tissue regeneration and immune responses (Spakova et al., 2021). Characterization techniques confirm their bioactivity, including the upregulation of collagen II and the downregulation of inflammatory cytokines, such as interleukin 1β (IL-1β) and tumor necrosis factor-α (TNF-α) (Otahal et al., 2021a).

Moreover, the anti-inflammatory properties of PRP have been highlighted in numerous investigations, suggesting that it can downregulate pro-inflammatory cytokines and modulate local inflammatory responses, which is crucial for effective healing (Bendinelli et al., 2010). PRP exerts anti-inflammatory effects in tendinopathy by suppressing nuclear factor kappa-light-chain-enhancer of activated B cells (NF-κB) signaling and TNF-α-induced interferon pathways, and its growth factors promote fibroblast chemotaxis, collagen synthesis, and tendon matrix repair, depending on preparation methods and patient-specific factors (Kale et al., 2024). EVs from blood products modulate gene expression in osteoarthritic chondrocytes, enhancing chondrogenic markers such as type II collagen and aggrecan, while also reducing inflammation by modulating cytokine release, including interleukin 6 (IL-6) and interleukin 8 (IL-8) (Otahal et al., 2020). PRP-derived exosomes alleviate knee OA by modulating the Wnt/β-catenin signaling pathway, promoting chondrogenic activity, reducing inflammation, and protecting chondrocytes from IL-1β-induced apoptosis and degeneration (Hou et al., 2023). Recent studies have shown that higher leukocyte content in PRP correlates with increased levels of pro-inflammatory mediators such as IL-1β and TNF-α, which may hinder cartilage repair; consequently, P-PRP has been reported as more effective than L-PRP in promoting cartilage regeneration (Jayaram et al., 2023).

Platelet-rich plasma has been applied in multiple fields, including plastic surgery, pediatric surgery, cardiac surgery, gynecology, urology, and ophthalmology (Gupta and Jain, 2024b). PRP is also used in tissue engineering as a carrier for cells and growth factors, often combined with biomaterials for bone, cartilage, skin, and soft-tissue regeneration. Although some studies report clinical benefits, others find minimal or no significant effects. A review of 50 studies found that evidence-based applications ranged from wound healing and cosmetics to burn care and fat grafting. However, inconsistent protocols and heterogeneity in study designs limit definitive conclusions, leaving their efficacy unproven in several medical fields (Hasiba-Pappas et al., 2022). In summary, the therapeutic mechanisms of PRP encompass a complex interplay between growth factors, cytokines, and EVs that collectively promote tissue repair, modulate inflammation, and enhance cellular regeneration. Extensive research backing its application in various medical domains underscores its significance as a promising regenerative therapy.

2.3 *Plasma Rich in Growth Factors*

Owing to challenges in standardizing PRP such as low handling efficiency, reliance on animal-derived thrombin for clotting, and inter-individual variability, Anitua et al. introduced plasma rich in growth factors (PRGF). This modified PRP protocol improves reproducibility by simplifying preparation and using $CaCl_2$ instead of thrombin to activate clotting (Masuki et al., 2016). The PRGF system eliminates leukocytes to reduce pro-inflammatory activity, employing a single-spin protocol, centrifuging sodium citrate-anticoagulated blood at 1850 rpm (580 g) for 8 min to yield three plasma fractions: an upper layer poor in growth factors, a middle layer with platelet concentrations comparable with peripheral blood, and a lower layer enriched with growth factors (Solakoglu et al., 2020). For in vitro studies, Anitua et al. activated PRGF with $CaCl_2$, incubated at 37 °C for 1 h, and the resulting supernatant was filtered and frozen at −80 °C for later use. For assays in vivo, PRGF was activated immediately before application to enable in situ scaffold formation (Anitua, et al., 2015a).

Content analysis of APBPs revealed that platelet concentrations were significantly elevated in both PRP and PRGF preparations, with PRGF showing substantially lower enrichment than PRP (2.84-fold versus 8.79-fold); notably, leukocytes were nearly eliminated in PRGF compared with PRP (0.015-fold versus 5.51-fold) (Masuki et al., 2016). Despite its name, PRGF contained markedly lower levels of growth factors (TGF-β1, PDGF, VEGF) and pro-inflammatory cytokines (IL-1β) than other PRP-derived products, suggesting that its therapeutic effects might be mediated by alternative molecular mechanisms (Masuki et al., 2016). For instance, PRGF promoted peripheral nerve regeneration by modulating macrophage polarization—shifting M1 to M2 phenotype—reducing pro-inflammatory cytokines (TNF-α, IL-1β, IL-6), enhancing axonal and Schwann cell markers, and supporting angiogenesis, ultimately improving nerve structure and muscle recovery in a rat model (Yadav et al., 2022). Intranasal administration of PRGF has shown promise in reversing neurodegeneration and enhancing memory in a transgenic mouse model of Alzheimer's disease. Moreover, PRGF exhibits neurotrophic activity in response to amyloid beta, supporting neuronal survival (Anitua et al., 2015b, 2021).

Plasma rich in growth factors is widely employed in ophthalmology as eye drops for treating ocular surface disorders. In 26 patients with ocular surface disease, PRGF combined with scleral contact lenses was safe, significantly reduced symptoms per dry eye symptom assessment scores, and was preferred over previous treatments by most patients, without affecting visual acuity (Wang et al., 2022). Beyond ophthalmic use, intratendinous PRGF injections provided significantly greater and more sustained improvements in pain and function than corticosteroids in chronic rotator cuff tendinopathy, with superior outcomes on three patient-reported scales at 6 and 12 months (Vaquerizo et al., 2023). Furthermore, PRGF-assisted nerve-release surgery led to significant long-term reductions in pain and functional improvement across carpal tunnel, recurrent carpal tunnel, and ulnar nerve entrapment syndromes of 39 participants, with no serious adverse events and consistent benefits at the 12-month follow-up (Galán et al., 2022). Moreover, in patients with frontal fibrosing alopecia, PRGF treatment was safe and effective in adjuvating standard care to halt hair loss, promote hair regrowth, reduce scalp inflammation, and improve symptom severity (García et al., 2023).

2.4 *Platelet Lysate*

To ensure the full release of platelet granule contents, platelet lysate (PL) was developed as an alternative to traditional PRP preparations. By inducing platelet rupture through physical or chemical methods, PL maximizes the availability of bioactive molecules, offering a more consistent and potent therapeutic profile. PL is produced by mechanically disrupting a platelet concentrate, typically PRP, through a freeze–thaw cycle that lyses platelets and releases growth factors stored in the α-granules of mature platelets, followed by centrifugation to remove cellular debris. Although no standardized preparation protocol exists, a commonly reported method involves 3–5 cycles of freezing PRP at −80 °C for 30 min and thawing at 37 °C, followed by centrifuging at 4000 rpm (1600 g) for 10 min to obtain the cell-free lysate (Brown et al., 2016). Compared with PRP, PL is less temperature-sensitive—enabling cryopreservation and long-term storage and may overcome the limitations of PRP related to donor variability, inconsistent growth-factor content, and incomplete platelet degranulation (Valtetsiotis et al., 2024).

Like PRP, PL contains key growth factors primarily IGF-1, PDGF isoforms (AA, BB, AB), TGF-β1, brain-derived neurotrophic factor (BDNF), EGF, VEGF, hepatocyte growth factor (HGF), and FGF, with some studies reporting higher concentrations of TGF-β1, PDGF-AB, IGF-1, and BDNF; additionally, PL includes plasma proteins (albumin, protease inhibitors such as α1-antitrypsin and α2-macroglobulin, and immunoglobulins G, A, and M), as well as minerals (sodium, potassium, magnesium, iron), vitamins, and glucose (da Fonseca et al., 2021). PL is primarily used as a serum-free supplement for the ex vivo expansion of mesenchymal stem cells (MSCs), serving as an effective substitute for fetal bovine serum (Henschler et al., 2019). Its utility extends to regenerative medicine, tissue engineering, and cell therapy owing to its content of soluble bioactive molecules and the absence of white blood cells and platelet membranes (Zamani et al., 2019).

PL has demonstrated therapeutic potential across a wide range of clinical and regenerative applications, including wound and ulcer healing, ocular graft-versus-host disease, OA, Parkinson's disease, tendon repair, alopecia, infertility, nerve and bone regeneration, oral mucositis, radicular pain, and musculoskeletal disorders (da Fonseca et al., 2021; Meftahpour et al., 2021). In a study of 50 women with knee OA, both PRP and PL intra-articular injections significantly improved pain and function compared with baseline, with PL showing higher growth factor concentrations than PRP, suggesting its potential as a promising therapeutic alternative (Hosseini et al., 2023). Despite these promising uses, few PL products have been specifically manufactured for direct clinical application, and controlled clinical trials are needed to establish therapeutic utility.

2.5 Platelet-Rich Fibrin

Fibrin is the final product of fibrinogen activation, playing a determining role in platelet aggregation during hemostasis. Various fibrin scaffolds have been developed to improve coagulation in bioengineering. Platelet-rich fibrin (PRF) is derived from whole blood without the use of anticoagulants immediately centrifuged at 2700 rpm, 750 g for 10 min, resulting in a leukocyte-rich fibrin matrix that releases platelet-derived growth factors more slowly and over a longer duration than PRP (Goswami et al., 2024; Karimi and Rockwell, 2019).

Centrifugation of whole blood without anticoagulants results in three layers: acellular, PPP on top, a middle PRF clot, and a bottom erythrocyte layer; the PRF clot, formed through rapid platelet activation and fibrin polymerization, traps platelets within its mesh and are extracted with sterile tweezers before refrigeration (Goswami et al., 2024). Although PRF is a platelet concentrate, it primarily functions as a fibrin-based technology, with the fibrin matrix alone contributing significantly to tissue healing; its slow polymerization produces a physiologically favorable architecture that enhances regenerative potential (Dohan et al., 2006). PRF has been clinically applied in both medical and dental procedures to improve wound healing and tissue regeneration, with successful use in treating facial soft-tissue defects, diabetic foot and venous leg ulcers, chronic leg wounds, midface volume loss, acne scars, urethrocutaneous fistulas, surgical scars, chronic rotator cuff tears, and acute traumatic eardrum perforations (Choukroun et al., 2017).

Furthermore, variants of the PRF process have been developed, improving its regenerative potential. Choukroun, a developer of PRF, later modified it to an advanced form of PRF (A-PRF), characterized by a softer fibrin clot and a higher concentration of leukocytes due to its lower-speed centrifugation protocol at 1300 rpm, 200 g for 8 min (Masuki et al., 2016). Quantification of growth factors in platelet concentrates found that A-PRF released PDGF-AB, TGF-β1, VEGF, and bone morphogenetic protein 2 (BMP-2) at 1–three-fold higher levels than other variations and consistently maintained them over 14 days (Lei et al., 2020). Gingival fibroblasts cultured with A-PRF showed significantly increased mRNA expression of PDGF, TGF-β, and collagen I after 3–7 days, indicating enhanced regenerative potential (Fujioka-Kobayashi et al., 2017).

Using even lower centrifugation speeds and shorter durations (700 rpm, 60 g for 3 min), a liquid formulation called injectable PRF (i-PRF) was developed, designed to be used alone or combined with biomaterials such as PRP but without the need for anticoagulants (Lei et al., 2020). Growth-factor analysis revealed that although PRP promoted higher early release, PRF sustained long-term delivery of PDGF-AA, PDGF-AB, EGF, and IGF-1 over 10 days; additionally, i-PRF significantly enhanced mRNA expression of PDGF at 3 days, TGF-β at 7 days, and collagen I at both time points compared with PRP (Pavlovic et al., 2021). I-PRF has been utilized in bone grafts to enhance stability and prevent granule migration and has also shown therapeutic potential in OA, temporomandibular joint disorders, and facial collagen synthesis protocols (Choukroun et al., 2017).

2.6 *Concentrated Growth Factors*

Concentrated growth factors (CGF), or growth factor concentrate (GFC), represent a modified PRF product obtained through alternating centrifugation speeds resulting in a rigid fibrin matrix rich in growth factors that has been shown to enhance the proliferation and osteogenic differentiation of periodontal ligament stem cells (PDLSCs) and bone marrow cells (Lei et al., 2020). Like PRF, CGF stimulates a sustained release of growth factors over 14 days; however, it demonstrates superior tensile strength, higher growth factor content, and greater long-term bone induction and tissue regeneration than PRP and PRF (Chen et al., 2023). CGF also lacks anticoagulants or bovine thrombin, overcoming a disadvantage of PRP preparation, and becoming an ideal biomaterial that promotes proliferation and differentiation in a 3D scaffold, supporting the organization and vascularization of regenerated cells (Li et al., 2021).

In the preparation of CGF, blood samples are collected in glass tubes and then processed by programmed centrifugation: 30 s of acceleration, 2 min at 2700 rpm (692 g), 4 min at 2400 rpm (547 g), 4 min at 2700 rpm (692 g), 3 min at 3000 rpm (855 g), and 30 s of deceleration and stop (Masuki et al., 2016). CGF preparation yields three main layers: an upper layer of PPP, a lower erythrocyte-rich layer, and a middle CGF gel layer composed of subfractions—a fibrin- and fibrillin-rich white upper portion supporting cell adhesion, a red lower portion containing erythrocytes, and a central buffy coat that can be isolated by scissoring and used either fresh or in freeze-dried form (Li et al., 2021). Fibrinolytic activation of platelets embedded in the CGF fibrin scaffold enables the sustained release of key growth factors—TGF-β1, PDGF-BB, IGF-1, BMP, VEGF, EGF, and FGF—that are essential for regulating stem-cell activity in tissue engineering (Li et al., 2021). CGF extract or conditioned medium is produced by incubating fresh or lyophilized CGF membranes in culture medium for 14 days or in inserts above cultured cells for 3 days, allowing diffusion of bioactive factors into the medium, which is then collected and supplemented for use in research (Tabatabaei et al., 2020).

Several studies have shown that CGF supplementation enhances dental pulp stem-cell (DPSC) proliferation in a dose-dependent manner and promotes the osteogenic and odontoblastic differentiation of DPSCs and stem cells from the apical papilla, as evidenced by mineralized nodule formation and upregulation of genes *COL1A1*, *ALP*, *OCN*, *DMP-1*, and *DSPP* (Jin et al., 2018). Similarly, CGF stimulates osteogenic differentiation in PDLSCs and bone-marrow-derived MSCs through increased expression of genes *COL1A1*, *ALP*, *OCN*, and *SP7* (Hong et al., 2018; Rochira et al., 2020). However, some studies report that highly concentrated CGF may inhibit cell proliferation owing to elevated levels of TGF-β and proteolytic enzymes (Tian et al., 2019; Yu & Wang, 2014).

Clinical trials have explored the use of CGF for various conditions, including peri-implantitis, guided tissue regeneration, and gingival recessions (Bozkurt Doȷan et al., 2015; Isler et al., 2018; Lei et al., 2020). In one study, a single injection of CGF on the right side of the face and PRP on the left in 60 patients showed that 66% of the 51 evaluated subjects had superior improvement with CGF, whereas only one favored PRP, indicating the effectiveness of CGF as a standalone treatment for facial folds (Sevilla et al., 2015). Intra-articular administration of CGF has been shown to effectively reduce pain and improve knee function in patients with OA (Gupta & Maffulli, 2024). CGF also demonstrates regenerative effects in ocular surface disorders, with optimization methods such as freezing and differential centrifugation enhancing therapeutic outcomes by preserving or modulating growth factor levels (Wessely-Szponder et al., 2021).

2.7 *Hyperacute Serum*

Hyperacute serum (HS), also known as serum from PRF, is a cell- and fibrin-free blood derivative extracted from PRF that promotes cell proliferation and differentiation (Simon et al., 2018). The preparation protocol of HS is similar to that of PRF; however, the serum is prepared by squeezing the serum from the formed PRF clot at the end of the procedure (Gupta & Aratikatla, 2024). Thus, HS is a liquid without coagulation factors that gives the final products a better cell proliferative effect with evidence for promoting hematopoiesis and adipogenic differentiation (Kardos et al., 2019; Simon et al., 2018).

Synovial fluid post-HS treatment analysis revealed significant protein-clustering patterns, suggesting coordinated biological responses (Olmos Calvo et al., 2021). Experimental comparisons between PRP and HS have indicated that HS consistently improves gene expression and matrix deposition in 3D chondrocyte models, demonstrating superior and more stable regenerative effects (Kuten-Pella et al., 2021). Further studies on EV-associated microRNAs (miRNAs) have revealed that PRP-derived EVs target NF-κB signaling, a key pathway in inflammation, and HS EVs primarily influence IL-6 and TGF-β/SMAD signaling, promoting tissue remodeling and immunomodulation (Otahal, et al., 2021b).

In co-culture models simulating inflammation, HS treatment markedly reduced key pro-inflammatory cytokines—including interleukins IL-2, IL-5, IL-7, IL-12, IL-15, and TNF-α—highlighting its anti-inflammatory potential (Calvo et al., 2021). Moreover, HS enhances the proliferative capacity of ischemia-damaged cells, exhibiting regenerative effects comparable with BMP-2 and bone morphogenetic protein 7 (BMP-7); however, it also contains antiangiogenic factors such as platelet factor 4 (PF-4), serpin, and tissue inhibitor of metalloproteinases-1 (TIMP-1), which may modulate its overall angiogenic potential (Vácz et al., 2018).

Hyperacute serum has shown promising results in enhancing cartilage-cell viability and reducing inflammation in OA. Clinical studies indicate that HS treatment leads to significant improvements in pain and function scores, with benefits lasting up to 6 months (Olmos Calvo et al., 2021). Another study suggests that HS administration might be safe and effective, increasing osteoarthritic chondrocyte viability, alleviating pain and inflammation, and improving function in patients with knee OA (Gupta & Aratikatla, 2024). Bains et al. found a significant correlation between HbA1c levels and the quantification of PRF and HS (Bains et al., 2023).

2.8 *Autologous Conditioned Serum*

Autologous conditioned serum (ACS) is a blood-derived product prepared by incubating whole blood with chromium(II) sulfate ($CrSO_4$)-coated glass beads and no anticoagulant at 37 °C for 24 h, followed by centrifugation at 3200 rpm (2790 g) for 10 min, resulting in serum enriched with anti-inflammatory cytokines such as IL-1 receptor antagonist (IL-1Ra), IL-4, IL-10, and IL-13 (Camargo Garbin & Morris, 2021; Linardi et al., 2019; Shakouri et al., 2022). Exposing blood to glass beads facilitates the synthesis of many cytokines that modulate inflammation and growth factors, including EGF, FGF, IGF-1, PDGF-ab, and TGF-β (Pishgahi et al., 2022; Wehling et al., 2007). ACS offers an advantage over PRP by specifically targeting inflammatory pathways through its enriched anti-inflammatory cytokines (Meijer et al., 2003). ACS outperforms PRP in pain relief, attributed to higher concentrations of PDGF-BB and IL-1Ra (Cheng et al., 2023).

Autologous conditioned serum has been used for local treatment of OA and radicular compression in both humans and horses, showing promise in managing tendinopathies, muscle injuries, and tunnel widening following anterior cruciate ligament reconstruction (Evans et al., 2016). A systematic review of ACS found that although in vitro studies showed inconsistent results, most clinical and observational studies reported beneficial effects; however, owing to the generally unclear to high risk of bias in these trials, its use in musculoskeletal lesions, though promising and safe, should still be approached with caution (Tokawa et al., 2022).

2.9 Autologous Protein Solution

Like ACS, autologous protein solution (APS) was developed to target inflammatory pathways mediated by IL-1β and TNF-α, with the idea of increasing its clinical use with a higher concentration of leukocytes and platelets (Camargo Garbin & Morris, 2021). APS is produced by first collecting L-PRP using an ACD-containing tube, centrifuging the blood in a separator at 2000 rpm (1744 g) for 15 min to isolate a platelet-rich solution, which is then transferred to a concentrator containing polyacrylamide beads and centrifuged again for 2 min to yield the final product (Camargo Garbin & Morris, 2021; Linardi et al., 2019). Contact between the blood and polyacrylamide beads stimulates the production of numerous cytokines and growth factors including soluble tumor necrosis factor receptor 1 (sTNF-RI), TGF-β, IL-1Ra, PDGF-AB/BB, IGF-1, and EGF (Camargo Garbin & Morris, 2021).

Razak et al. reviewed in vitro, preclinical, and clinical studies on single intra-articular injections of APS for early-stage OA, finding that 2–3 mL of APS produced from 50 to 60 mL of blood consistently elevated anti-inflammatory cytokines—including IL-1ra, soluble IL-1 receptor II (sIL-1RII), IL-1β, and tumor necrosis factor receptor 2 (sTNF-RII)—and led to significant improvements in pain reduction and joint function in patients with knee OA (Razak et al., 2021). In vitro, Woodell-May J et al. found that pre-incubation of chondrocytes with APS inhibited MMP-13 synthesis before human recombinant IL-1β or TNF-α addition (Woodell-May et al., 2011). Moreover, Matuska et al. observed that APS prevented cartilage matrix degradation more effectively than recombinant antagonists (Matuska et al., 2013). One of the clinical studies found that intra-articular APS was safe and led to significant pain relief up to 3 years after a single injection in patients with mild to moderate knee OA, with better outcomes in those with less cartilage loss regardless of baseline pain (Kon et al., 2020). Furthermore, Van Genechten et al. reported a 54% response rate at 12 months following one or two APS injections in middle-aged women with advanced patellofemoral cartilage wear, noting reduced efficacy of the second dose in initially poor responders after 3 months, whereas in baseline MRI-detected major synovitis better pain relief and functional improvement were predicted (Van Genechten et al., 2021). Overall, these findings support APS as a safe and effective therapeutic option that modulates inflammation, protects cartilage, and provides durable pain relief and functional improvement in patients with OA, particularly when administered early and in those with less advanced joint damage.

2.10 Enhancing Blood Products with Natural Compounds

Although some have proposed ozone therapy as an enhancement to blood-based treatments, it remains prohibited by the FDA owing to safety concerns and a lack of reliable evidence. In contrast, safer and biologically grounded approaches—such as integrating dietary bioactive compounds—offer promising strategies to enhance autologous therapies, particularly through nutritional modulation. Table 4.2

Table 4.2 Selected compounds investigated for enhancing autologous peripheral blood-based therapies

Compound	Condition	Strategy	Relevant outcomes	Reference
EGCG and quercetin	Hemostasis disturbed by HOCl	PRP was obtained by centrifugation of whole blood at 200 × *g* for 20 min	Preincubation (1 min) with quercetin (5–25 μM) or EGCG (10–25 μM) abolished HOCl-evoked inhibition of platelet aggregation	Misztal et al. (2022)
Resveratrol	Hemostasis in high-risk cardiac patients with aspirin resistance	PRP was isolated from aspirin-sensitive and aspirin-resistant and adjusted to 275 to 325 × 10^3 platelets/μL	Maximal aggregation of aspirin-resistant platelets to collagen in PRP was significantly decreased by resveratrol (10^{-5} mol/L)	Stef et al. (2006)
	Hemostasis in acute hyperglycemia (dilution and incubation of PRP with 25 mM glucose)	PRP was obtained by centrifugation of whole blood within 2 h of collection at 100 × g for 20 min	Resveratrol decreased procoagulant platelet number and attenuated oxygen consumption rate in PRP	Gauer et al. (2024)
Curcumin	Rat model of acute sciatic nerve injury	Four treatments were randomly allocated: (1) nerve repair only; (2) nerve repair + local PRP; (3) nerve repair + intraperitoneal curcumin; and (4) nerve repair + local PRP + intraperitoneal curcumin	The combined administration of curcumin and PRP as adjuvants to nerve repair enhanced axonal regeneration in terms of electrophysiological and histological parameters	Zavala et al. (2023)
	Rat model of diabetic wounds	Four treatments were randomly allocated: (1) 500 μL saline control; (2) 500 μL PRP injection; (3) 2.5 × 10^5 nonpreconditioned hASCs +500 μL PRP; (4) and 2.5 × 10^5 curcumin-preconditioned hASCs +500 μL PRP	Curcumin-preconditioned hASCs combined with PRP enhanced wound healing by improving cell viability under glucose stress and promoting faster closure, neovascularization, reduced inflammation, and upregulation of key regenerative markers at the wound site	Ghufran et al. (2020)
Carbogenic nanodrug from curcumin	Chronic wounds in diabetes	This study proposes an autologous platelet concentrates-based synergistic platform (CurCDs@iPRF-MA) composed of i-PRF, GelMA, and CurCDs	This study proposes CurCDs@iPRF-MA as a therapeutic platform for diabetic-wound healing, given its potential to restore mitochondrial homeostasis, activate OXPHOS, modulate macrophages and ROS, and enhance vascularization through autologous growth factor release	Ren et al. (2025)
Inorganic polyP	Cell culture and murine models of wound healing	Inorganic polyP-depleted keratinocyte and fibroblast cell culture models and a murine model of wound healing were used to determine if polyP, PRP, or combined treatment could accelerate wound healing	In vitro, PRP and polyP reduced the scratch-wound area, although only PRP enhanced cell proliferation, whereas in vivo, PRP accelerated re-epithelialization, and polyP synergistically enhanced this effect by promoting longer epithelial tongues than PRP alone	Carney et al. (2020)

(continued)

Table 4.2 (continued)

Compound	Condition	Strategy	Relevant outcomes	Reference
Tannic acid	Scaffolds for wound healing	PRF was obtained by centrifugation of whole blood at 700 × g for 10 min, added to form the gel, and then tannic acid with different concentrations (0.5%, 1%, 5%, and 10%) was applied for crosslinking	The 10% tannic acid scaffold showed a threefold increase in mechanical strength, reducing swelling (from 49% to 5%) and degradation (from 63% to 8%), improving the physical properties of PRF without negatively affecting its biological properties	Haghparast-Kenarsari et al. (2024)
Vitamin C	PTRCTs	110 patients with PTRCTs were randomly allocated to two groups: (1) 1.5 ml saline +1.5 ml PRP; and (2) 1.5 ml vitamin C + 1.5 ml PRP; and functional scores and pain reduction were used to evaluate the outcomes at 3 months	Both PRP injection alone and PRP combined with vitamin C led to significant reductions in pain and enhancements in function scores over time, but further research is required to ascertain if the combination therapy offers statistically significant advantages	Mohammadivahedi et al. (2024)
	eqASCs	Treatments of eqASCs included: (1) control group (2) vitamin C; (3) PRP; (4) vitamin C + PRP; vitamin C and PRP were used when indicated at 0.1 mM and 5% (v/v) respectively	Combining vitamin C and PRP positively affected the ability of eqASCs to differentiate in vitro into mesodermal lineages during 14 days of culture	Castro et al. (2014)
Vitamin C and vitamin E	Nonhealing DFUs	Twenty-five patients with nonhealing DFUs were treated with PRP–FG dressing plus oral vitamin E and C (intervention group) or PRP-FG dressing plus placebo (control group) for 8 weeks	In patients with nonhealing DFU, PRP-FG dressings combined with oral vitamins E and C significantly improved wound healing and reduced oxidative stress and inflammation markers compared with PRP-FG alone	Yarahmadi et al. (2021)
Vitamin D	Male Sprague–Dawley rat models of bone defects	Fifty-two male rats divided into six groups: (1) no material; (2) chitosan gel; (3) 0.05 g PRP + calcitriol; (4) chitosan-poloxamer gel +0.025 g of PRP; (5) chitosan-poloxamer gel + calcitriol; (6) chitosan-poloxamer gel +0.025 g of PRP + calcitriol; preparations contained 0.025 g of chitosan (5%)-poloxamer 407 (15%) and 150 μg calcitriol in 30 μl of ethanol	A formulation of active vitamin D (calcitriol), PRP, and chitosan significantly enhanced bone defect repair, as evidenced by improved bone mineral density, defect shrinkage with new cell colonization, and stronger expression of RANK, FGF23, and VDR compared to controls	Turgut et al. (2023)
	Osteogenic differentiation and proliferation of hMSCs	Three different treatments of hMSCs included: (1) different concentrations of PRP; (2) different PRP concentrations +1,25(OH)$_2$D$_3$ (50 nM); and (3) serial co-treatment with PRP-1.0× + 1,25(OH)$_2$D$_3$ (50 nM)	PRP alone inhibited osteogenic differentiation and proliferation of hMSCs but co-treatment with vitamin D synergistically stimulated osteogenic differentiation	Feng et al. (2010)

(continued)

Table 4.2 (continued)

Compound	Condition	Strategy	Relevant outcomes	Reference
HA	Knee OA	Ten RCTs involving 943 patients were analyzed to evaluate whether a combination of PRP and HA is more effective and safer than injection alone for treating knee OA	PRP + HA treatment resulted in enhanced efficacy when compared with HA monotherapy, and presented lower incidence of adverse events than either HA or PRP alone	Gao et al. (2024)

CurCD carbogenic nanodrug from curcumin, *DFU* diabetic foot ulcer, *EGCG* epigallocatechin gallate, *eqASC* equine adipose mesenchymal stem cell, *FG* fibrin glue, *GelMA* gelatin methacryloyl, *HA* hyaluronic acid, *hASC* human adipose-derived cell, *hMSC* human mesenchymal stem cell, *HOCl* hypochlorous acid, *OA* osteoarthritis, *OXPHOS* oxidative phosphorylation, *polyP* polyphosphate, *PRP* platelet-rich plasma, *PTRCT* partial-thickness rotator cuff tear, *ROS* reactive oxygen species

summarizes selected compounds explored as APBP enhancers, including the medical conditions modeled, descriptions of study methodologies, and the key therapeutic outcomes observed.

In addition to nutritional factors, metabolic modulators such as phosphoenolpyruvate increase red-blood-cell levels of 2,3-BPG, improving oxygen delivery and capillary perfusion in ischemic conditions (Zhou et al., 2019). Hypoxia preconditioned plasma, obtained by culturing anticoagulated blood under physiological temperature and low oxygen, alters plasma properties to enhance its therapeutic effects (Hadjipanayi & Schilling, 2014), possibly by increasing 2,3-BPG levels.

Together, these biologically compatible enhancements reflect a shift toward integrative, patient-specific strategies in regenerative medicine. By leveraging the natural synergy between blood components and targeted nutritional or metabolic interventions, clinicians may improve therapeutic efficacy, reduce oxidative stress, and optimize the preparation methods of APBPs.

3 Safety, Ethical, and Regulatory Considerations

The use of blood-derived therapies presents ethical and regulatory challenges owing to a lack of standardized preparation and treatment protocols, which results in variable outcomes and limits reproducibility across clinical settings. These inconsistencies, combined with low-quality evidence from small, underpowered, and biased studies, complicate the interpretation and generalization of clinical findings.

One of the most pressing challenges is the high degree of variability in PRP and other autologous formulations, particularly in platelet concentration and growth-factor content. Many clinical studies are methodologically flawed, or lack appropriate controls, contributing to a weak evidence base that limits confidence in clinical efficacy. A systematic review by Chahla et al. found that among 105 orthopedic studies using PRP, only 11 fully detailed their preparation protocols for reproducibility, and just 17 reported qualitative metrics on PRP composition, highlighting a significant lack of standardization and transparency (Chahla et al., 2017). Without rigorous study designs and standardized reporting, drawing meaningful conclusions or comparing results across trials is unattainable. To bridge these gaps, future research must prioritize well-designed, large-scale clinical trials that clearly define therapeutic parameters, endpoints, and patient selection criteria (Hasiba-Pappas et al., 2022; Lana et al., 2021; Martinez-Zapata et al., 2016; Napit et al., 2023).

Commercialization has introduced additional issues related to transparency and quality control. Industry-sponsored studies of APBPs and the equipment used in their production often exhibit bias and weaker methodological standards, underscoring the need for independent, well-powered clinical trials to validate therapeutic efficacy (Nesello et al., 2020).

Safety concerns also persist, particularly in vulnerable populations such as patients with immunodeficiency, where the effects of APBPs are not fully understood. Unregulated blood therapies at unlicensed facilities pose serious risks of HIV and other bloodborne infections owing to inadequate infection control and lack of proper client records, as evidenced by an HIV outbreak linked to cosmetic PRP microneedling facials in New Mexico (Stadelman-Behar et al., 2024). Ethical practice requires careful patient selection, transparent communication about risks and benefits, and rigorous monitoring of treatment outcomes.

Although PRP is exempt from FDA drug classification because of its autologous nature, it lacks formal FDA approval, limiting insurance coverage and regulatory oversight, despite its widespread clinical use (Beitzel et al., 2015). This regulatory gray area emphasizes the need for standardized preparation protocols, robust clinical trials, and the careful, responsible promotion of PRP therapies to ensure patient safety and maintain clinical credibility.

4 Future Perspectives and Research Gaps

The future of APBPs hinges on improving standardization, expanding therapeutic applications, and integrating emerging technologies. One of the central research priorities is the establishment of consistent preparation and administration protocols to reduce variability in clinical outcomes and enhance reproducibility across studies and patient populations. Standardization will be essential for regulatory acceptance, broader clinical use, and effective comparison between blood products.

Recent investigations have demonstrated that lyophilized HS formulations can retain high cell viability, matching or surpassing liquid HS and PRP (Calvo et al., 2021). The capability of lyophilized preparations to maintain biological activity over time presents an advantage in product stability and clinical usability, particularly in settings where cold-chain logistics are limited.

Similarly, cryopreservation techniques are being optimized to extend the shelf-life of therapies such as PRP. Studies have shown that freezing PRP at −20 °C for up to 3 months not only preserves the biological integrity of the product but can also enhance the concentration of critical proteins such as PDGF-BB and EGF (Rodriguez et al., 2020). Subsequent centrifugation allows modulation of specific growth factors such as TGF-β1, offering further opportunities for tailoring treatments to specific clinical needs (Rodriguez et al., 2020). These approaches are particularly relevant for managing chronic conditions, such as ocular surface disorders, where long-term, stable formulations are necessary.

Finally, although this chapter focuses on APBPs, a One-Health approach also encourages the study of xenogeneic blood-derived therapies, including those sourced from veterinary medicine, which may offer valuable insights and translational potential across species. Human placental hydrolysate, which contains HGF, EGF, and FGF, has been evaluated as a potential strategy for treating canine liver failure, which shares similar molecular and cellular mechanisms with human liver disease. In an in vitro study, this hydrolysate promoted the long-term culture of hepatocyte-like cells derived from canine bone marrow (Neo et al., 2020). Moreover, it has been shown to induce liver regeneration in rats, HepG2 cells, and primary human hepatocytes (Lee et al., 2019).

The potential applications of blood therapies are broadening to include not only regenerative medicine and musculoskeletal repair but also infectious diseases and chronic inflammatory conditions. Through the combined efforts of clinical researchers, regulatory bodies, and biomedical engineers, the field of autologous blood therapies is poised to make significant contributions to precision medicine, improving both patient outcomes and the sustainability of health care delivery.

5 Conclusion

This chapter explores the evolving landscape of autologous blood-derived therapies, emphasizing their growing importance in modern medical science. APBPs, including PRP, PRF, and ACS, have demonstrated significant therapeutic potential in various clinical contexts, particularly in wound healing, musculoskeletal injuries, and degenerative joint diseases. These therapies exert their effects by modulating cell-signaling pathways, regulating inflammatory cytokines, promoting angiogenesis, and stimulating tissue regeneration.

By harnessing the patient's blood components, autologous therapies offer several advantages over traditional pharmacological or surgical interventions. Chief among these is the reduction in immunogenicity and risk of adverse reactions, which enhances both safety and biocompatibility. Furthermore, the minimally invasive nature of these treatments supports their use in outpatient settings and for long-term disease management.

Nonetheless, the clinical translation of these therapies remains hindered by key challenges, particularly the lack of standardized protocols for product preparation, variability in therapeutic content, and inconsistencies in outcome reporting. The current body of evidence is often limited by small sample sizes, heterogeneity in study design, and methodological flaws, which collectively restrict the generalizability of findings. Addressing these limitations through rigorous, multicenter clinical trials and standardized methodological frameworks will be essential to validate efficacy and enable widespread clinical adoption.

Looking ahead, innovations in synthetic biology and bioengineering are poised to enhance the precision, delivery, and scalability of blood-based therapies. Future developments in standardized preparation protocols could enable the production of blood-derived products from a single donor source, potentially overcoming supply constraints and improving consistency across treatments. Together, these technologies may expand the clinical reach of blood products and refine their integration into patient care.

In summary, autologous blood therapies represent a compelling convergence of natural healing mechanisms and biomedical innovation. As the field matures, these interventions have the potential to become a cornerstone of next-generation regenerative medicine—delivering safe, effective, and personalized treatments for a wide range of acute and chronic conditions. Ensuring equitable access, robust evidence, and ethical application will be critical in realizing their global impact.

References

Anitua, E., Pelacho, B., Prado, R., & Aguirre, J. J. (2015a). Infiltration of plasma rich in growth factors enhances in vivo angiogenesis and improves reperfusion and tissue remodeling after severe hind limb ischemia. *Journal of Controlled Release, 202*, 31–39. https://doi.org/10.1016/j.jconrel.2015.01.029

Anitua, E., Prado, R., Azkargorta, M., Rodriguez-Suárez, E., Iloro, I., Casado-Vela, J., Elortza, F., & Orive, G. (2015b). High-throughput proteomic characterization of plasma rich in growth factors (PRGF-Endoret)-derived fibrin clot interactome. *Journal of Tissue Engineering and Regenerative Medicine, 9*(11), E1–E12. https://doi.org/10.1002/TERM.1721

Anitua, E., Muruzabal, F., Pino, A., Prado, R., Azkargorta, M., Elortza, F., & Merayo-lloves, J. (2021). Proteomic characterization of plasma rich in growth factors and undiluted autologous serum. *International Journal of Molecular Sciences, 22*(22). https://doi.org/10.3390/IJMS222212176

Bains, V., Mahendra, J., Mahendra, L., Mittal, M., & Gunam, V. (2023). Quantitative association of platelet-rich fibrin (PRF) and hyperacute serum (HAS) with glycemic control (HbA1c) in chronic periodontitis patients: An ex vivo study. *Journal of Pharmacy and Bioallied Sciences, 15*(5), 601–607. https://doi.org/10.4103/JPBS.JPBS_428_22

Beitzel, K., Allen, D., Apostolakos, J., Russell, R. P., McCarthy, M. B., et al. (2015). US definitions, current use, and FDA stance on use of platelet-rich plasma in sports medicine. *The Journal of Knee Surgery, 28*(1), 29–34. https://doi.org/10.1055/S-0034-1390030

Bendinelli, P., Matteucci, E., Dogliotti, G., Corsi, M. M., Banfi, G., Maroni, P., & Desiderio, M. A. (2010). Molecular basis of anti-inflammatory action of platelet-rich plasma on human chondrocytes: Mechanisms of NF-κB inhibition via HGF. *Journal of Cellular Physiology, 225*(3), 757–766. https://doi.org/10.1002/JCP.22274

Bowers, E., Slaughter, A., Frenette, P. S., Kuick, R., Pello, O. M., & Lucas, D. (2018). Granulocyte-derived TNFα promotes vascular and hematopoietic regeneration in the bone marrow. *Nature Medicine, 24*(1), 95–102. https://doi.org/10.1038/NM.4448

Bozkurt Doɀan, Ş., Öngöz Dede, F., Balli, U., Atalay, E. N., & Durmuşlar, M. C. (2015). Concentrated growth factor in the treatment of adjacent multiple gingival recessions: A split-mouth randomized clinical trial. *Journal of Clinical Periodontology, 42*(9), 868–875. https://doi.org/10.1111/JCPE.12444

Brewer, D. D., & Brewer, D. D. (2014). A systematic review of Autohemotherapy as a treatment for Urticaria and eczema. *Cureus, 6*(12). https://doi.org/10.7759/CUREUS.233

Brown, M. N., Shiple, B. J., & Scarpone, M. (2016). Regenerative approaches to tendon and ligament conditions. *Physical Medicine and Rehabilitation Clinics of North America, 27*(4), 941–984. https://doi.org/10.1016/j.pmr.2016.07.003

Cabezas-Wallscheid, N., Klimmeck, D., Hansson, J., Lipka, D. B., Reyes, A., Wang, Q., Weichenhan, D., Lier, A., Von Paleske, L., Renders, S., Wünsche, P., Zeisberger, P., Brocks, D., Gu, L., Herrmann, C., Haas, S., Essers, M. A. G., Brors, B., Eils, R., et al. (2014). Identification of regulatory networks in HSCs and their immediate progeny via integrated proteome, transcriptome, and DNA methylome analysis. *Cell Stem Cell, 15*(4), 507–522. https://doi.org/10.1016/j.stem.2014.07.005

Calvo, I. O., Kuten-Pella, O., Kramer, K., Madár, Á., Takács, S., Kardos, D., Simon, D., Erdö-Bonyár, S., Berki, T., De Luna, A., Nehrer, S., & Lacza, Z. (2021). Optimization of lyophilized hyperacute serum (has) as a regenerative therapeutic in osteoarthritis. *International Journal of Molecular Sciences, 22*(14), 7496. https://doi.org/10.3390/IJMS22147496/S1

Camargo Garbin, L., & Morris, M. J. (2021). A comparative review of autologous conditioned serum and autologous protein solution for treatment of osteoarthritis in horses. *Frontiers in Veterinary Science, 8*, 602978. https://doi.org/10.3389/FVETS.2021.602978

Carney, B. C., Simbulan-Rosenthal, C. M., Gaur, A., Browne, B. J., Moghe, M., Crooke, E., Moffatt, L. T., Shupp, J. W., & Rosenthal, D. S. (2020). Inorganic polyphosphate in platelet rich plasma accelerates re-epithelialization in vitro and in vivo. *Regenerative Therapy, 15*, 138–148. https://doi.org/10.1016/j.reth.2020.07.004

Castro, F. O., Torres, A., Cabezas, J., & Rodríguez-Alvarez, L. L. (2014). Combined use of platelet rich plasma and vitamin C positively affects differentiation in vitro to mesodermal lineage of adult adipose equine mesenchymal stem cells. *Research in Veterinary Science, 96*(1), 95–101. https://doi.org/10.1016/j.rvsc.2013.12.005

Chahla, J., Cinque, M. E., Piuzzi, N. S., Mannava, S., Geeslin, A. G., Murray, I. R., Dornan, G. J., Muschler, G. F., & LaPrade, R. F. (2017). A call for standardization in platelet-rich plasma preparation protocols and composition reporting: A systematic review of the clinical Orthopaedic literature. *Journal of Bone and Joint Surgery American, 99*(20), 1769–1779. https://doi.org/10.2106/JBJS.16.01374

Chen, L., Cheng, J., Cai, Y., Zhang, J., Yin, X., & Luan, Q. (2023). Efficacy of concentrated growth factor (CGF) in the surgical treatment of oral diseases: A systematic review and meta-analysis. *BMC Oral Health, 23*(1), 712. https://doi.org/10.1186/S12903-023-03357-5

Cheng, P. G., Yang, K. D., Huang, L. G., Wang, C. H., & Ko, W. S. (2023). Comparisons of cytokines, growth factors and clinical efficacy between platelet-rich plasma and autologous conditioned serum for knee osteoarthritis management. *Biomolecules, 13*(3). https://doi.org/10.3390/BIOM13030555

Choukroun, J., Aalam, A. A., & Miron, R. J. (2017). *Platelet rich fibrin "PRF" and regenerative medicine: 'The Low-speed concept'* (pp. 21–42). Humana Press. https://doi.org/10.1007/978-3-319-55645-1_2

Coillard, A., & Segura, E. (2019). In vivo differentiation of human monocytes. *Frontiers in Immunology, 10*(AUG), 469933. https://doi.org/10.3389/FIMMU.2019.01907/BIBTEX

Collins, T., Alexander, D., & Barkatali, B. (2021). Platelet-rich plasma: A narrative review. *EFORT Open Reviews, 6*(4), 225. https://doi.org/10.1302/2058-5241.6.200017

da Fonseca, L., Santos, G. S., Huber, S. C., Setti, T. M., Setti, T., & Lana, J. F. (2021). Human platelet lysate – A potent (and overlooked) orthobiologic. *Journal of Clinical Orthopaedics and Trauma, 21*, 101534. https://doi.org/10.1016/J.JCOT.2021.101534

Dash, S. P., Gupta, S., & Sarangi, P. P. (2024). Monocytes and macrophages: Origin, homing, differentiation, and functionality during inflammation. *Heliyon, 10*(8), e29686.

Davis, V. L., Abukabda, A. B., Radio, N. M., Witt-Enderby, P. A., Clafshenkel, W. P., Cairone, J. V., & Rutkowski, J. L. (2014). Platelet-rich preparations to improve healing. Part II: Platelet activation and enrichment, leukocyte inclusion, and other selection criteria. *Journal of Oral Implantology, 40*(4), 511–521. https://doi.org/10.1563/AAID-JOI-D-12-00106

Delong, J. M., Russell, R. P., & Mazzocca, A. D. (2012). Platelet-rich plasma: The PAW classification system. *Arthroscopy Journal of Arthroscopic and Related Surgery, 28*(7), 998–1009. https://doi.org/10.1016/J.ARTHRO.2012.04.148

Dhurat, R., & Sukesh, M. (2014). Principles and methods of preparation of platelet-rich plasma: A review and author's perspective. *Journal of Cutaneous and Aesthetic Surgery, 7*(4), 189. https://doi.org/10.4103/0974-2077.150734

Di Matteo, B., Filardo, G., Kon, E., & Marcacci, M. (2015). Platelet-rich plasma: Evidence for the treatment of patellar and Achilles tendinopathy – A systematic review. *Musculoskeletal Surgery, 99*(1). https://doi.org/10.1007/S12306-014-0340-1

Do Amaral, R. J. F. C., Da Silva, N. P., Haddad, N. F., Lopes, L. S., Ferreira, F. D., Filho, R. B., Cappelletti, P. A., De Mello, W., Cordeiro-Spinetti, E., & Balduino, A. (2016). Platelet-rich plasma obtained with different anticoagulants and their effect on platelet numbers and mesenchymal stromal cells behavior in vitro. *Stem Cells International, 2016*. https://doi.org/10.1155/2016/7414036

Dohan Ehrenfest, D. M., Andia, I., Zumstein, M. A., Zhang, C. Q., Pinto, N. R., & Bielecki, T. (2014). Classification of platelet concentrates (platelet-rich plasma-PRP, platelet-rich fibrin-PRF) for topical and infiltrative use in orthopedic and sports medicine: Current consensus, clinical implications and perspectives. *Muscles, Ligaments and Tendons Journal, 4*(1), 3. https://doi.org/10.32098/mltj.01.2014.02

Dohan, D. M., Choukroun, J., Diss, A., Dohan, S. L., Dohan, A. J. J., Mouhyi, J., & Gogly, B. (2006). Platelet-rich fibrin (PRF): A second-generation platelet concentrate. Part I: Technological concepts and evolution. *Oral Surgery, Oral Medicine, Oral Pathology, Oral Radiology, and Endodontology, 101*(3). https://doi.org/10.1016/J.TRIPLEO.2005.07.008

Dwyer, D. F., Barrett, N. A., Austen, K. F., Kim, E. Y., Brenner, M. B., Shaw, L., Yu, B., Goldrath, A., Mostafavi, S., Regev, A., Rhoades, A., Moodley, D., Yoshida, H., Mathis, D., Benoist, C., Nabekura, T., Lam, V., Lanier, L. L., Brown, B., et al. (2016). Expression profiling of constitutive mast cells reveals a unique identity within the immune system. *Nature Immunology, 17*(7), 878–887. https://doi.org/10.1038/NI.3445

Evans, C. H., Chevalier, X., & Wehling, P. (2016). Autologous conditioned serum. *Physical Medicine and Rehabilitation Clinics of North America, 27*(4), 893–908. https://doi.org/10.1016/j.pmr.2016.06.003

Feng, Y., Sun, Y., Jia, W., & Zhang, C. (2010). Platelet-rich plasma and 1,25(OH)2 vitamin D3 synergistically stimulate osteogenic differentiation of adult human mesenchymal stem cells. *Biotechnology Letters, 32*(5), 635–642. https://doi.org/10.1007/S10529-009-0198-8

Foster, T. E., Puskas, B. L., Mandelbaum, B. R., Gerhardt, M. B., & Rodeo, S. A. (2009). Platelet-rich plasma: From basic science to clinical applications. *American Journal of Sports Medicine, 37*(11), 2259–2272. https://doi.org/10.1177/0363546509349921

Fujioka-Kobayashi, M., Miron, R. J., Hernandez, M., Kandalam, U., Zhang, Y., & Choukroun, J. (2017). Optimized platelet-rich fibrin with the low-speed concept: Growth factor release, biocompatibility, and cellular response. *Journal of Periodontology, 88*(1), 112–121. https://doi.org/10.1902/JOP.2016.160443

Galán, V., Iñigo-Dendariarena, I., Galán, I., Prado, R., Padilla, S., & Anitua, E. (2022). The effectiveness of plasma rich in growth factors (PRGF) in the treatment of nerve compression syndromes of the upper extremity: A retrospective observational clinical study. *Journal of Clinical Medicine, 11*(16). https://doi.org/10.3390/JCM11164789

Gao, J., Ma, Y., Tang, J., Zhang, J., & Zuo, J. (2024). Efficacy and safety of platelet-rich plasma and hyaluronic acid combination therapy for knee osteoarthritis: A systematic review and meta-analysis. *Archives of Orthopaedic and Trauma Surgery, 144*(9). https://doi.org/10.1007/S00402-024-05442-Y

García, A., Navarro, M. R., Ramirez, A., Pino, A., Navarro, A., Moles, I., Gallego, E., & Anitua, E. (2023). Plasma rich in growth factors as an adjuvant treatment for the Management of Frontal Fibrosing Alopecia: A retrospective observational clinical study. *Journal of Cutaneous Medicine and Surgery, 27*(4), 340–349. https://doi.org/10.1177/12034754231177599

Gauer, J. S., Ajanel, A., Kaselampao, L. M., Candir, I., MacCannell, A. D. V., Roberts, L. D., Campbell, R. A., & Ariëns, R. A. S. (2024). Plant-derived compounds normalize platelet bioenergetics and function in hyperglycemia. *Research and Practice in Thrombosis and Haemostasis, 8*(6). https://doi.org/10.1016/j.rpth.2024.102548

Gharpinde, M. R., Pundkar, A., Shrivastava, S., Patel, H., & Chandanwale, R. (2024). A comprehensive review of platelet-rich plasma and its emerging role in accelerating bone healing. *Cureus, 16*(2). https://doi.org/10.7759/CUREUS.54122

Ghufran, H., Mehmood, A., Azam, M., Butt, H., Ramzan, A., Yousaf, M. A., Ejaz, A., Tarar, M. N., & Riazuddin, S. (2020). Curcumin preconditioned human adipose derived stem cells co-transplanted with platelet rich plasma improve wound healing in diabetic rats. *Life Sciences, 257*. https://doi.org/10.1016/J.LFS.2020.118091

Glenn, A., & Armstrong, C. E. (2019). Physiology of red and white blood cells. *Anaesthesia and Intensive Care Medicine, 20*(3), 170–174. https://doi.org/10.1016/j.mpaic.2019.01.001

Goswami, P., Chaudhary, V., Arya, A., Verma, R., Vijayakumar, G., & Bhavani, M. (2024). Platelet-rich fibrin (PRF) and its application in dentistry: A literature review. *Journal of Pharmacy and Bioallied Sciences, 16*(Suppl 1), S5–S7. https://doi.org/10.4103/JPBS.JPBS_435_23

Guo, S. C., Tao, S. C., Yin, W. J., Qi, X., Yuan, T., & Zhang, C. Q. (2017). Exosomes derived from platelet-rich plasma promote the re-epithelization of chronic cutaneous wounds via activation of YAP in a diabetic rat model. *Theranostics, 7*(1), 81–96. https://doi.org/10.7150/THNO.16803

Gupta, A., & Aratikatla, A. (2024). Hyperacute serum and knee osteoarthritis. *Cureus, 16*(1). https://doi.org/10.7759/CUREUS.53118

Gupta, A., & Jain, V. (2024a). Autologous conditioned plasma (ACP) and osteoarthritis of the knee: A review of current clinical evidence. *Cureus, 16*(1), e52693. https://doi.org/10.7759/CUREUS.52693

Gupta, A., & Jain, V. K. (2024b). Autologous peripheral blood-derived orthobiologics: Different types and their effectiveness in managing knee osteoarthritis. *World Journal of Orthopedics, 15*(5), 400. https://doi.org/10.5312/WJO.V15.I5.400

Gupta, A., & Maffulli, N. (2024). Growth factor concentrate (GFC) for the Management of Osteoarthritis of the knee: A systematic review. *Indian Journal of Orthopaedics, 58*(7), 829–834. https://doi.org/10.1007/S43465-024-01172-W

Hadjipanayi, E., & Schilling, A. F. (2014). Regeneration through autologous hypoxia preconditioned plasma. *Organogenesis, 10*(2), 164. https://doi.org/10.4161/ORG.29208

Haghparast-Kenarsari, T., Abdouss, M., Heidari Keshel, S., Heshmatipour, Z., Rahimi, A., & Biazar, E. (2024). Improving properties of platelet-rich fibrin scaffold with tannic acid for wound healing. *International Wound Journal, 21*(4). https://doi.org/10.1111/IWJ.14571

Hasiba-Pappas, S. K., Tuca, A. C., Luze, H., Nischwitz, S. P., Zrim, R., Geißler, J. C. J., Lumenta, D. B., Kamolz, L. P., & Winter, R. (2022). Platelet-rich plasma in plastic surgery: A systematic review. *Transfusion Medicine and Hemotherapy, 49*(3), 129–142. https://doi.org/10.1159/000524353

Henschler, R., Gabriel, C., Schallmoser, K., Burnouf, T., & Koh, M. B. C. (2019). Human platelet lysate current standards and future developments. *Transfusion, 59*(4), 1407–1413. https://doi.org/10.1111/TRF.15174

Hoeffel, G., & Ginhoux, F. (2018). Fetal monocytes and the origins of tissue-resident macrophages. *Cellular Immunology, 330*, 5–15. https://doi.org/10.1016/j.cellimm.2018.01.001

Hong, S., Chen, W., & Jiang, B. (2018). A comparative evaluation of concentrated growth factor and platelet-rich fibrin on the proliferation, migration, and differentiation of human stem cells of the apical papilla. *Journal of Endodontics, 44*(6), 977–983. https://doi.org/10.1016/J.JOEN.2018.03.006

Hosseini, S., Soltani-Zangbar, M. S., Zamani, M., Yaghoubi, Y., Rikhtegar Ghiasi, R., Motavalli, R., Ghassabi, A., Iranzad, R., Mehdizadeh, A., Shakouri, S. K., Pishgahi, A., & Yousefi, M. (2023). Comparative evaluation of autologous platelet-rich plasma and platelet lysate in patients with knee osteoarthritis. *Growth Factors, 41*(3), 165–177. https://doi.org/10.1080/08977194.2023.2227273

Hou, Y., Wen, X., Zhou, L., & Fang, X. (2023). The value of platelet-rich plasma-derived extracellular vesicles in modern medicine. *Annals of Medicine, 55*(2). https://doi.org/10.1080/07853890.2023.2287705

Isler, S. C., Soysal, F., Ceyhanlı, T., Bakırarar, B., & Unsal, B. (2018). Regenerative surgical treatment of peri-implantitis using either a collagen membrane or concentrated growth factor: A 12-month randomized clinical trial. *Clinical Implant Dentistry and Related Research, 20*(5), 703–712. https://doi.org/10.1111/CID.12661

Italiani, P., & Boraschi, D. (2014). From monocytes to M1/M2 macrophages: Phenotypical vs. functional differentiation. *Frontiers in Immunology, 5*(OCT). https://doi.org/10.3389/FIMMU.2014.00514

Italiani, P., & Boraschi, D. (2015). New insights into tissue macrophages: From their origin to the development of memory. *Immune Network, 15*(4), 167. https://doi.org/10.4110/IN.2015.15.4.167

Jayaram, P., Mitchell, P. J. T., Shybut, T. B., Moseley, B. J., & Lee, B. (2023). Leukocyte-rich platelet-rich plasma is predominantly anti-inflammatory compared with leukocyte-poor platelet-rich plasma in patients with mild-moderate knee osteoarthritis: A prospective, descriptive laboratory study. *American Journal of Sports Medicine, 51*(8), 2133–2140. https://doi.org/10.1177/03635465231170394

Jin, R., Song, G., Chai, J., Gou, X., Yuan, G., & Chen, Z. (2018). Effects of concentrated growth factor on proliferation, migration, and differentiation of human dental pulp stem cells in vitro. *Journal of Tissue Engineering, 9*, 2041731418817505. https://doi.org/10.1177/2041731418817505

Kale, P., Patel, H., & Jaiswal, A. M. (2024). Mechanisms, efficacy, and clinical applications of platelet-rich plasma in tendinopathy: A comprehensive review. *Cureus, 16*(7). https://doi.org/10.7759/CUREUS.65636

Kardos, D., Simon, M., Vácz, G., Hinsenkamp, A., Holczer, T., Cseh, D., Sárközi, A., Szenthe, K., Bánáti, F., Szathmary, S., Nehrer, S., Kuten, O., Masteling, M., Lacza, Z., & Hornyák, I. (2019). The composition of hyperacute serum and platelet-rich plasma is markedly different despite the similar production method. *International Journal of Molecular Sciences, 20*(3). https://doi.org/10.3390/IJMS20030721

Karimi, K., & Rockwell, H. (2019). The benefits of platelet-rich fibrin. *Facial Plastic Surgery Clinics of North America, 27*(3), 331–340. https://doi.org/10.1016/j.fsc.2019.03.005

Klion, A. D., Ackerman, S. J., & Bochner, B. S. (2020). Contributions of eosinophils to human health and disease. *Annual Review of Pathology: Mechanisms of Disease, 15*, 179–209. https://doi.org/10.1146/ANNUREV-PATHMECHDIS-012419-032756

Kon, E., Engebretsen, L., Verdonk, P., Nehrer, S., & Filardo, G. (2020). Autologous protein solution injections for the treatment of knee osteoarthritis: 3-year results. *American Journal of Sports Medicine, 48*(11), 2703–2710. https://doi.org/10.1177/0363546520944891

Kuhn, V., Diederich, L., Keller, T. C. S., Kramer, C. M., Lückstädt, W., Panknin, C., Suvorava, T., Isakson, B. E., Kelm, M., & Cortese-Krott, M. M. (2017). Red blood cell function and dysfunction: Redox regulation, nitric oxide metabolism, anemia. *Antioxidants and Redox Signaling, 26*(13), 718–742. https://doi.org/10.1089/ARS.2016.6954

Kuten-Pella, O., De Luna, A., Kramer, K., Neubauer, M., Nehrer, S., & Lacza, Z. (2021). Regenerative potential of blood-derived products in 3d osteoarthritic chondrocyte culture system. *Current Issues in Molecular Biology, 43*(2), 665–675. https://doi.org/10.3390/CIMB43020048

Lana, J. F. S. D., da Fonseca, L. F., Macedo, R. D. R., Mosaner, T., Murrell, W., Kumar, A., Purita, J., & de Andrade, M. A. P. (2021). Platelet-rich plasma vs bone marrow aspirate concentrate: An overview of mechanisms of action and orthobiologic synergistic effects. *World Journal of Stem Cells, 13*(2), 155. https://doi.org/10.4252/WJSC.V13.I2.155

Lee, T. H., Park, D. S., Jang, J. Y., Lee, I., Kim, J. M., Choi, G. S., Oh, C. T., Kim, J. Y., Han, H. J., Han, B. S., & Joh, J. W. (2019). Human placenta hydrolysate promotes liver regeneration via activation of the cytokine/growth factor-mediated pathway and anti-oxidative effect. *Biological and Pharmaceutical Bulletin, 42*(4), 607–616. https://doi.org/10.1248/BPB.B18-00712

Lei, L., Yu, Y., Han, J., Shi, D., Sun, W., Zhang, D., & Chen, L. (2020). Quantification of growth factors in advanced platelet-rich fibrin and concentrated growth factors and their clinical efficacy as adjunctive to the GTR procedure in periodontal intrabony defects. *Journal of Periodontology, 91*(4), 462–472. https://doi.org/10.1002/JPER.19-0290

Li, Z., Liu, L., Wang, L., & Song, D. (2021). The effects and potential applications of concentrated growth factor in dentin–pulp complex regeneration. *Stem Cell Research and Therapy, 12*(1), 1–10. https://doi.org/10.1186/S13287-021-02446-Y/FIGURES/1

Linardi, R. L., Dodson, M. E., Moss, K. L., King, W. J., & Ortved, K. F. (2019). The effect of autologous protein solution on the inflammatory Cascade in stimulated equine chondrocytes. *Frontiers in Veterinary Science, 6*(MAR), 64. https://doi.org/10.3389/FVETS.2019.00064

Mace, E. M. (2023). Human natural killer cells: Form, function, and development. *Journal of Allergy and Clinical Immunology, 151*(2), 371–385. https://doi.org/10.1016/J.JACI.2022.09.022

Manole, C. G., Soare, C., Ceafalan, L. C., & Voiculescu, V. M. (2024). Platelet-rich plasma in dermatology: New insights on the cellular mechanism of skin repair and regeneration. *Life, 14*(1). https://doi.org/10.3390/LIFE14010040

Martin, C., Burdon, P. C. E., Bridger, G., Gutierrez-Ramos, J. C., Williams, T. J., & Rankin, S. M. (2003). Chemokines acting via CXCR2 and CXCR4 control the release of neutrophils from the bone marrow and their return following senescence. *Immunity, 19*(4), 583–593. https://doi.org/10.1016/S1074-7613(03)00263-2

Martinez-Zapata, M. J., Martí-Carvajal, A. J., Solà, I., Expósito, J. A., Bolíbar, I., Rodríguez, L., Garcia, J., & Zaror, C. (2016). Autologous platelet-rich plasma for treating chronic wounds. *Cochrane Database of Systematic Reviews, 2016*(5). https://doi.org/10.1002/14651858.CD006899.PUB3

Masuki, H., Okudera, T., Watanebe, T., Suzuki, M., Nishiyama, K., Okudera, H., Nakata, K., Uematsu, K., Su, C.-Y., & Kawase, T. (2016). Growth factor and pro-inflammatory cytokine contents in platelet-rich plasma (PRP), plasma rich in growth factors (PRGF), advanced platelet-rich fibrin (A-PRF), and concentrated growth factors (CGF). *International Journal of Implant Dentistry, 2*(1). https://doi.org/10.1186/S40729-016-0052-4

Matuska, A., O'Shaughnessey, K., King, W., & Woodell-May, J. (2013). Autologous solution protects bovine cartilage explants from IL-1α- and TNFα-induced cartilage degradation. *Journal of Orthopaedic Research, 31*(12), 1929–1935. https://doi.org/10.1002/JOR.22464

Mayoral Rojals, V., Amescua Garcia, C., Denegri, P., Narvaez Tamayo, M. A., & Varrassi, G. (2023). The invasive management of pain: Diagnosis and new treatment options. *Cureus, 15*(7). https://doi.org/10.7759/CUREUS.42717

Meftahpour, V., Malekghasemi, S., Baghbanzadeh, A., Aghebati-Maleki, A., Pourakbari, R., Fotouhi, A., & Aghebati-Maleki, L. (2021). Platelet lysate: A promising candidate in regenerative medicine. *Regenerative Medicine, 16*(1), 71–85. https://doi.org/10.2217/RME-2020-0065

Meijer, H., Reinecke, J., Becker, C., Tholen, G., & Wehling, P. (2003). The production of anti-inflammatory cytokines in whole blood by physico-chemical induction. *Inflammation Research, 52*(10), 404–407. https://doi.org/10.1007/S00011-003-1197-1

Minetti, G., Migliaccio, A. R., & Fibach, E. (2021). Editorial: Membrane processes in erythroid development and red cell life time. *Frontiers in Physiology, 12*, 655117. https://doi.org/10.3389/FPHYS.2021.655117

Misztal, T., Golaszewska, A., Marcińczyk, N., Tomasiak-łozowska, M., Szymanowska, M., Chabielska, E., & Rusak, T. (2022). Natural polyphenols may normalize hypochlorous acid-evoked hemostatic abnormalities in human blood. *Antioxidants, 11*(4). https://doi.org/10.3390/ANTIOX11040779

Mohammadivahedi, F., Sadeghifar, A., Farsinejad, A., Jambarsang, S., & Mirhosseini, H. (2024). Comparative efficacy of platelet-rich plasma (PRP) injection versus PRP combined with vitamin C injection for partial-thickness rotator cuff tears: A randomized controlled trial. *Journal of Orthopaedic Surgery and Research, 19*(1). https://doi.org/10.1186/S13018-024-04917-3

Moras, M., Lefevre, S. D., & Ostuni, M. A. (2017). From erythroblasts to mature red blood cells: Organelle clearance in mammals. *Frontiers in Physiology, 8*(DEC). https://doi.org/10.3389/FPHYS.2017.01076

Murray, I. R., McAdams, T. R., Hammond, K. E., Haddad, F. S., Rodeo, S. A., Abrams, G. D., Bankston, L., Bedi, A., Boublik, M., Bowen, M., Bradley, J. P., Cooper, D. E., Craythorne, C., Curl, L. A., ElAttrache, N., Gazzaniga, D. S., Kaplan, K., Khalfayan, E. E., Larson, C., et al. (2023). The use of biologics in NFL athletes: An expert consensus of NFL team physicians. *Orthopaedic Journal of Sports Medicine, 11*(2). https://doi.org/10.1177/23259671221143778

Napit, I. B., Shrestha, D., Neupane, K., Adhikari, A., Dhital, R., Koirala, R., Gopali, L., Ilozumba, O., Gill, P., Watson, S. I., Choudhury, S., & Lilford, R. J. (2023). Autologous blood products: Leucocyte and platelets rich fibrin (L-PRF) and platelets rich plasma (PRP) gel to promote cutaneous ulcer healing – A systematic review. *BMJ Open, 13*(12). https://doi.org/10.1136/BMJOPEN-2023-073209

Neo, S., Makiishi, E., Fujimoto, A., & Hisasue, M. (2020). Human placental hydrolysate promotes the long-term culture of hepatocyte-like cells derived from canine bone marrow. *Journal of Veterinary Medical Science, 82*(12), 1821–1827. https://doi.org/10.1292/JVMS.20-0320

Nesello, P. F. T., Baroni, A. C., & Selistre, L. D. S. (2020). Level of evidence and industry sponsorship associated with favorable outcomes in publications on platelet-rich-plasma therapy in musculoskeletal disorders. *Revista Brasileira de Ortopedia, 55*(3), 263–268. https://doi.org/10.1055/S-0039-1700834

Noetzli, L. J., French, S. L., & Machlus, K. R. (2019). New insights into the differentiation of megakaryocytes from hematopoietic progenitors. *Arteriosclerosis, Thrombosis, and Vascular Biology, 39*(7), 1288–1300. https://doi.org/10.1161/ATVBAHA.119.312129

Nussbaum, J. C., Van Dyken, S. J., Von Moltke, J., Cheng, L. E., Mohapatra, A., Molofsky, A. B., Thornton, E. E., Krummel, M. F., Chawla, A., Liang, H. E., & Locksley, R. M. (2013). Type 2 innate lymphoid cells control eosinophil homeostasis. *Nature, 502*(7470), 245–248. https://doi.org/10.1038/NATURE12526

Olmos Calvo, I., Fodor, E., Kardos, D., Hornyák, I., Hinsenkamp, A., Kuten-Pella, O., Gyevnár, Z., Erdélyi, G., Bárdos, T., Paukovits, T. M., Magos, K., Béres, G., Nehrer, S., & Lacza, Z. (2021). A pilot clinical study of hyperacute serum treatment in osteoarthritic knee joint: Cytokine changes and clinical effects. *Current Issues in Molecular Biology, 43*(2), 637–649. https://doi.org/10.3390/CIMB43020046/S1

Oneto, P., Zubiry, P. R., Schattner, M., & Etulain, J. (2020). Anticoagulants interfere with the Angiogenic and regenerative responses mediated by platelets. *Frontiers in Bioengineering and Biotechnology, 8*. https://doi.org/10.3389/FBIOE.2020.00223

Otahal, A., Kramer, K., Kuten-Pella, O., Weiss, R., Stotter, C., Lacza, Z., Weber, V., Nehrer, S., & De Luna, A. (2020). Characterization and Chondroprotective effects of extracellular vesicles from plasma- and serum-based autologous blood-derived products for osteoarthritis therapy. *Frontiers in Bioengineering and Biotechnology, 8*. https://doi.org/10.3389/FBIOE.2020.584050

Otahal, A., Kramer, K., Kuten-pella, O., Moser, L. B., Neubauer, M., Lacza, Z., Nehrer, S., & De Luna, A. (2021a). Effects of extracellular vesicles from blood-derived products on osteoarthritic chondrocytes within an inflammation model. *International Journal of Molecular Sciences, 22*(13). https://doi.org/10.3390/IJMS22137224

Otahal, A., Kuten-Pella, O., Kramer, K., Neubauer, M., Lacza, Z., Nehrer, S., & De Luna, A. (2021b). Functional repertoire of EV-associated miRNA profiles after lipoprotein depletion via ultracentrifugation and size exclusion chromatography from autologous blood products. *Scientific Reports, 11*(1). https://doi.org/10.1038/S41598-021-84234-5

Patel, H., Pundkar, A., Shrivastava, S., Chandanwale, R., & Jaiswal, A. M. (2023). A comprehensive review on platelet-rich plasma activation: A key player in accelerating skin wound healing. *Cureus, 15*(11). https://doi.org/10.7759/CUREUS.48943

Pavlovic, V., Ciric, M., Jovanovic, V., Trandafilovic, M., & Stojanovic, P. (2021). Platelet-rich fibrin: Basics of biological actions and protocol modifications. *Open Medicine (Poland), 16*(1), 446–454. https://doi.org/10.1515/MED-2021-0259

Pietras, E. M., Reynaud, D., Kang, Y. A., Carlin, D., Calero-Nieto, F. J., Leavitt, A. D., Stuart, J. A., Göttgens, B., & Passegué, E. (2015). Functionally distinct subsets of lineage-biased multipotent progenitors control blood production in normal and regenerative conditions. *Cell Stem Cell, 17*(1), 35–46. https://doi.org/10.1016/j.stem.2015.05.003

Pishgahi, A., Zamani, M., Mehdizadeh, A., Roshangar, L., Afkham-Daghdaghan, M., Pourabbas, B., & Yousefi, M. (2022). The therapeutic effects of autologous conditioned serum on knee osteoarthritis: An animal model. *BMC Research Notes, 15*(1), 1–7. https://doi.org/10.1186/S13104-022-06166-1/FIGURES/3

Pretorius, J., Habash, M., Ghobrial, B., Alnajjar, R., & Ellanti, P. (2023). Current status and advancements in platelet-rich plasma therapy. *Cureus, 15*(10). https://doi.org/10.7759/CUREUS.47176

Quicke, J. G., Conaghan, P. G., Corp, N., & Peat, G. (2022). Osteoarthritis year in review 2021: Epidemiology & therapy. *Osteoarthritis and Cartilage, 30*(2), 196–206. https://doi.org/10.1016/j.joca.2021.10.003

Ragni, E., Taiana, M. M., Čengić, T., de Girolamo, L., & Ostojić, M. (2025). PRP or not PRP: Is the debate surrounding platelets-based blood-derived products evolving? *Knee Surgery, Sports Traumatology, Arthroscopy, 33*(5). https://doi.org/10.1002/KSA.12655

Razak, H. R. B. A., Chew, D., Kazezian, Z., & Bull, A. M. J. (2021). Autologous protein solution: A promising solution for osteoarthritis? *EFORT Open Reviews, 6*(9), 716–726. https://doi.org/10.1302/2058-5241.6.200040

Ren, S., Lv, H., Chen, S., Zhou, J., Chen, S., Chen, J., Luo, J., Guo, Y., Wang, H., Zhai, J., & Zhou, Y. (2025). Photoresponsive blood-derived protein hydrogels packed with bioactive carbon dots modulate mitochondrial homeostasis and reprogram metabolism for chronic wound healing in diabetes. *ACS Applied Materials and Interfaces*. https://doi.org/10.1021/ACSAMI.5C00635/SUPPL_FILE/AM5C00635_SI_001.PDF

Rochira, A., Siculella, L., Damiano, F., Palermo, A., Ferrante, F., Carluccio, M. A., Calabriso, N., Giannotti, L., & Stanca, E. (2020). Concentrated growth factors (CGF) induce osteogenic differentiation in human bone marrow stem cells. *Biology, 9*(11), 370. https://doi.org/10.3390/BIOLOGY9110370

Rodriguez, A. E., Gisbert, S., Palazón, A., & Alio, J. L. (2020). Quantification of growth factors and fibronectin in diverse preparations of platelet-rich plasma for the treatment of ocular surface disorders (E-PRP). *Translational Vision Science and Technology, 9*(6), 1–10. https://doi.org/10.1167/TVST.9.6.22

Sevilla, G., Dhurat, R., Shetty, G., Kadam, P., & Totey, S. (2015). Safety and efficacy of growth factor concentrate in the treatment of nasolabial fold correction: Split face pilot study. *Indian Journal of Dermatology, 60*(5), 520. https://doi.org/10.4103/0019-5154.159628

Shah, H., Eisenbarth, S., Tormey, C. A., & Siddon, A. J. (2021). Behind the scenes with basophils: An emerging therapeutic target. *Immunotherapy Advances, 1*(1). https://doi.org/10.1093/IMMADV/LTAB008

Shakouri, S. K., Dolati, S., Santhakumar, J., Thakor, A. S., & Yarani, R. (2022). Autologous conditioned serum for degenerative diseases and prospects. *Growth Factors, 39*(1–6), 59–70. https://doi.org/10.1080/08977194.2021.2012467

Simon, M., Major, B., Vácz, G., Kuten, O., Hornyák, I., Hinsenkamp, A., Kardos, D., Bagó, M., Cseh, D., Sárközi, A., Horvathy, D., Nehrer, S., & Lacza, Z. (2018). The effects of hyperacute serum on the elements of the human subchondral bone marrow niche. *Stem Cells International, 2018*, 4854619. https://doi.org/10.1155/2018/4854619

Solakoglu, Ö., Heydecke, G., Amiri, N., & Anitua, E. (2020). The use of plasma rich in growth factors (PRGF) in guided tissue regeneration and guided bone regeneration. A review of his-

tological, immunohistochemical, histomorphometrical, radiological and clinical results in humans. *Annals of Anatomy Anatomischer Anzeiger, 231*, 151528. https://doi.org/10.1016/J.AANAT.2020.151528
Spakova, T., Janockova, J., & Rosocha, J. (2021). Characterization and therapeutic use of extracellular vesicles derived from platelets. *International Journal of Molecular Sciences, 22*(18). https://doi.org/10.3390/IJMS22189701
Stadelman-Behar, A. M., Gehre, M. N., Atallah, L., Clarke, T., Leonso, A.-A., Jojola, F., Zheng, H., Jia, H., Lyss, S. B., Switzer, W. M., Grytdal, S. P., Durham, M., Salas, N. M., Sievers, M., & Smelser, C. (2024). Investigation of presumptive HIV transmission associated with receipt of platelet-rich plasma microneedling facials at a Spa among former Spa clients – New Mexico, 2018–2023. *MMWR. Morbidity and Mortality Weekly Report, 73*(16), 372–376. https://doi.org/10.15585/MMWR.MM7316A3
Stef, G., Csiszar, A., Lerea, K., Ungvari, Z., & Veress, G. (2006). Resveratrol inhibits aggregation of platelets from high-risk cardiac patients with aspirin resistance. *Journal of Cardiovascular Pharmacology, 48*(2), 1–5. https://doi.org/10.1097/01.FJC.0000238592.67191.AB
Summers, C., Rankin, S. M., Condliffe, A. M., Singh, N., Peters, A. M., & Chilvers, E. R. (2010). Neutrophil kinetics in health and disease. *Trends in Immunology, 31*(8), 318–324. https://doi.org/10.1016/j.it.2010.05.006
Tabatabaei, F., Aghamohammadi, Z., & Tayebi, L. (2020). In vitro and in vivo effects of concentrated growth factor on cells and tissues. *Journal of Biomedical Materials Research - Part A, 108*(6), 1338–1350. https://doi.org/10.1002/JBM.A.36906
Tao, S. C., Yuan, T., Rui, B. Y., Zhu, Z. Z., Guo, S. C., & Zhang, C. Q. (2017). Exosomes derived from human platelet-rich plasma prevent apoptosis induced by glucocorticoid-associated endoplasmic reticulum stress in rat osteonecrosis of the femoral head via the Akt/Bad/Bcl-2 signal pathway. *Theranostics, 7*(3), 733–750. https://doi.org/10.7150/THNO.17450
Thomas, M. R., & Storey, R. F. (2015). The role of platelets in inflammation. *Thrombosis and Haemostasis, 114*(3), 449–458. https://doi.org/10.1160/TH14-12-1067
Tian, S., Wang, J., Dong, F., Du, N., Li, W., Song, P., & Liu, Y. (2019). Concentrated growth factor promotes dental pulp cells proliferation and mineralization and facilitates recovery of dental pulp tissue. *Medical Science Monitor: International Medical Journal of Experimental and Clinical Research, 25*, 10016. https://doi.org/10.12659/MSM.919316
Tokawa, P. K. A., Brossi, P. M., & Baccarin, R. Y. A. (2022). Autologous conditioned serum in equine and human orthopedic therapy: A systematic review. *Research in Veterinary Science, 146*, 34–52. https://doi.org/10.1016/j.rvsc.2022.03.009
Turgut, C. T., Ozturk, C., Kotil, T., Gungor, S., Ozsoy, Y., Dirican, A., Okyar, A., & Yaltirik, M. (2023). The effect of biocompatible carriers with vitamin D and platelet-rich plasma (PRP) on bone defects. *Hippokratia, 27*(2), 48. https://pmc.ncbi.nlm.nih.gov/articles/PMC11268316/
Vácz, G., Major, B., Gaál, D., Petrik, L., Horváthy, D. B., Han, W., Holczer, T., Simon, M., Muir, J. M., Hornyák, I., & Lacza, Z. (2018). Hyperacute serum has markedly better regenerative efficacy than platelet-rich plasma in a human bone oxygen-glucose deprivation model. *Regenerative Medicine, 13*(5), 531–543. https://doi.org/10.2217/RME-2017-0141
Valent, P., Klion, A. D., Horny, H. P., Roufosse, F., Gotlib, J., Weller, P. F., Hellmann, A., Metzgeroth, G., Leiferman, K. M., Arock, M., Butterfield, J. H., Sperr, W. R., Sotlar, K., Vandenberghe, P., Haferlach, T., Simon, H. U., Reiter, A., & Gleich, G. J. (2012). Contemporary consensus proposal on criteria and classification of eosinophilic disorders and related syndromes. *Journal of Allergy and Clinical Immunology, 130*(3). https://doi.org/10.1016/j.jaci.2012.02.019
Valtetsiotis, K., Di Martino, A., Brunello, M., D'Agostino, C., Poluzzi, R., Ferri, R., Mora, P., Traina, F., & Faldini, C. (2024). Platelet lysate for the treatment of osteoarthritis: A systematic review of preclinical and clinical studies. *Musculoskeletal Surgery, 108*(3), 275. https://doi.org/10.1007/S12306-024-00827-Z
Van Genechten, W., Vuylsteke, K., Swinnen, L., Martinez, P. R., & Verdonk, P. (2021). Autologous protein solution as selective treatment for advanced patellofemoral osteoarthritis in the middle-

aged female patient: 54% response rate at 1 year follow-up. *Knee Surgery, Sports Traumatology, Arthroscopy, 29*(3), 988–997. https://doi.org/10.1007/S00167-020-06064-8

Vaquerizo, V., García-López, M., Mena-Rosón, A., Prado, R., Padilla, S., & Anitua, E. (2023). Plasma rich in growth factors versus corticosteroid injections for management of chronic rotator cuff tendinopathy: A prospective double-blind randomized controlled trial with 1 year of follow-up. *Journal of Shoulder and Elbow Surgery, 32*(3), 555–564. https://doi.org/10.1016/j.jse.2022.08.017

Wang, M., Yennam, S., McMillin, J., Chen, H., et al. (2022). Combined therapy of ocular surface disease with plasma rich in growth factors and scleral contact lenses. *Ocular Surface, 23*, 162–168. https://doi.org/10.1016/j.jtos.2021.09.003

Wehling, P., Moser, C., et al. (2007). Autologous conditioned serum in the treatment of orthopedic diseases: The Orthokine® therapy. *BioDrugs, 21*(5), 323–332. https://doi.org/10.2165/00063030-200721050-00004

Wessely-Szponder, J., Zdziennicka, J., Junkuszew, A., Latalski, M., Świeca, M., & Szponder, T. (2021). Prospects and applications of natural blood-derived products in regenerative medicine. *International Journal of Molecular Sciences, 23*(1), 472. https://doi.org/10.3390/IJMS23010472

Woodell-May, J., Matuska, A., Oyster, M., Welch, Z., O'Shaughnessey, K., & Hoeppner, J. (2011). Autologous protein solution inhibits MMP-13 production by IL-1β and TNFα-stimulated human articular chondrocytes. *Journal of Orthopaedic Research, 29*(9), 1320–1326. https://doi.org/10.1002/JOR.21384

Wu, Y., & Hirschi, K. K. (2021). Tissue-resident macrophage development and function. *Frontiers in Cell and Developmental Biology, 8*. https://doi.org/10.3389/FCELL.2020.617879

Wu, J., Piao, Y., Liu, Q., & Yang, X. (2021). Platelet-rich plasma-derived extracellular vesicles: A superior alternative in regenerative medicine? *Cell Proliferation, 54*(12). https://doi.org/10.1111/CPR.13123

Yadav, A., Ramasamy, T. S., Lin, S. C., Chen, S. H., Lu, J., Liu, Y. H., Lu, F. I., Hsueh, Y. Y., Lin, S. P., & Wu, C. C. (2022). Autologous platelet-rich growth factor reduces M1 macrophages and modulates inflammatory microenvironments to promote sciatic nerve regeneration. *Biomedicine, 10*(8). https://doi.org/10.3390/BIOMEDICINES10081991

Yarahmadi, A., Saeed Modaghegh, M. H., Mostafavi-Pour, Z., Azarpira, N., Mousavian, A., Bonakdaran, S., Jarahi, L., Samadi, A., Peimani, M., & Hamidi Alamdari, D. (2021). The effect of platelet-rich plasma-fibrin glue dressing in combination with oral vitamin E and C for treatment of non-healing diabetic foot ulcers: A randomized, double-blind, parallel-group, clinical trial. *Expert Opinion on Biological Therapy, 21*(5), 687–696. https://doi.org/10.1080/14712598.2021.1897100

Yu, B., & Wang, Z. (2014). Effect of concentrated growth factors on beagle periodontal ligament stem cells in vitro. *Molecular Medicine Reports, 9*(1), 235–242. https://doi.org/10.3892/MMR.2013.1756

Yun, S. H., Sim, E. H., Goh, R. Y., Park, J. I., & Han, J. Y. (2016). Platelet activation: The mechanisms and potential biomarkers. *BioMed Research International, 2016*. https://doi.org/10.1155/2016/9060143

Zamani, M., Yaghoubi, Y., Movassaghpour, A., Shakouri, K., Mehdizadeh, A., Pishgahi, A., & Yousefi, M. (2019). Novel therapeutic approaches in utilizing platelet lysate in regenerative medicine: Are we ready for clinical use? *Journal of Cellular Physiology, 234*(10), 17172–17186. https://doi.org/10.1002/JCP.28496

Zavala, A., Martinez, P. C., Gutierrez, G. G., Vara, M. D., & Pawlikowski, W. D. (2023). The combined use of curcumin and platelet-rich plasma enhances axonal regeneration in acute nerve injuries: An experimental study in a rat model. *Journal of Hand and Microsurgery, 15*(1), 31–36. https://doi.org/10.1055/S-0040-1721562

Zhang, X., Lin, Y., Xin, J., Zhang, Y., Yang, K., Luo, Y., & Wang, B. (2024). Red blood cells in biology and translational medicine: Natural vehicle inspires new biomedical applications. *Theranostics, 14*(1), 220–248. https://doi.org/10.7150/THNO.87425

Zhou, S., Giannetto, M., DeCourcey, J., Kang, H., Kang, N., Li, Y., Zheng, S., Zhao, H., Simmons, W. R., Wei, H. S., Bodine, D. M., Low, P. S., Nedergaard, M., & Wan, J. (2019). Oxygen tension-mediated erythrocyte membrane interactions regulate cerebral capillary hyperemia. *Science Advances, 5*(5). https://doi.org/10.1126/SCIADV.AAW4466

Chapter 5
Nutritional Intervention in Human Therapy

Hernández Tobías Esther Alhelí and Vázquez-Rodríguez Jesús Alberto

1 Introduction

The 2030 Agenda for Sustainable Development, adopted by the United Nations in 2015, provides a global framework to end poverty, protect the environment, and ensure health and well-being for all. Among its 17 Sustainable Development Goals, three are closely linked to nutrition and health, such as SDG 2 (Zero Hunger), which addresses the eradication of hunger and all forms of malnutrition; SDG 3 (Good Health and Well-being), which supports evidence-based interventions aimed at prevention of noncommunicable diseases and the improvement of overall health across the life course; and SDG 12 (Responsible Consumption and Production), which promotes sustainable food systems development and dietary patterns that align with both human health and environmental sustainability, emphasizing minimization of the ecological footprint of diets. As a cross-cutting determinant, nutrition plays a dual role in both shaping and enabling progress toward sustainable health, equity, and environmental resilience.

Therefore, dietary therapies must evolve beyond individual treatment paradigms to align with broader global objectives. As the burden of noncommunicable diseases rises, so does the need for nutritional strategies that are preventive, sustainable, and equitable. Aligning these interventions with the Sustainable Development Goals ensures a multidimensional impact: improving individual health, reducing health disparities, and minimizing environmental footprints. By integrating principles of food equity, climate resilience, and public-health nutrition, dietary therapies can become powerful tools, not only in clinical care but also in advancing a healthier, more sustainable future.

H. T. E. Alhelí (✉) · V.-R. J. Alberto
Facultad de Salud Pública y Nutrición, Universidad Autónoma de Nuevo León, San Nicolás de los Garza, Mexico
e-mail: esther.hernandeztb@uanl.edu.mx

M. S. Heya (ed.), *Natural Alternatives and Nanotechnology Applied to the One Health Approach*, https://doi.org/10.1007/978-3-032-03675-9_5

2 Global Nutrition

Current trends in nutrition reveal a complex landscape where overconsumption of ultra-processed, energy-dense foods coexists with persistent micronutrient deficiencies and rising rates of diet-related noncommunicable diseases (NCDs) (Vilar-Compte et al., 2021). This nutritional imbalance is not only a key driver of obesity, diabetes, cardiovascular disease (CVD), and certain cancers but also contributes to mental-health disorders such as anxiety and depression, especially when diets are low in essential fatty acids and fiber. The double burden of malnutrition, in which undernutrition occurs with overnutrition, depicts the complex landscape of global nutrition and challenges due to the uneven outcomes across diverse populations. These disparities are influenced not only by individual choices but for several drivers such as geography, socioeconomic status, and education that shape dietary habits and hence the health outcomes (Kim et al., 2022).

Moreover, unsustainable dietary patterns characterized by excessive intake of saturated fat, added sugars, and industrially processed foods exert a double burden by harming public health and accelerating environmental degradation through increased greenhouse gas emissions, water use, and loss of biodiversity. These dynamics highlight the urgent need for nutrition strategies that are both health-promoting and environmentally sustainable.

3 Nutritional Therapies

To effectively reduce the global burden of NCDs through sustainable dietary strategies, a range of nutritional interventions have been proposed. However, this remains a highly complex challenge, as standardized dietary models often fail to account for the interindividual variability that characterizes chronic conditions. Such variability is partially rooted in genetic architecture but is also shaped by sociocultural, educational, and socioeconomic determinants across diverse populations. In response to this complexity, current research is increasingly focused on personalized and systemic approaches, such as plant-based dietary patterns, traditional diets as sustainable alternatives, and precision nutrition informed by multi-omic technologies. These strategies are aimed not only at optimizing individual metabolic responses but also at aligning with planetary health goals, offering a more integrative framework for addressing the dual challenges of chronic-disease prevention and sustainable development (Klapp et al., 2022; Lane et al., 2022; Loughman et al., 2021).

4 Plant-Based Diets, Chronic Disease Prevention, and Environmental Sustainability

Plant-based diets, characterized by high consumption of fruits, vegetables, whole grains, legumes, nuts, and seeds, with minimal to no animal-source foods; have generated increasing scientific and public interest over the past decade. Concurrently, the world faces mounting challenges: escalating rates of NCDs (e.g., CVD, type 2 diabetes, cancer), rising pharmacological resistance, pharmaceutical environmental pollution, and intensifying human and zoonotic disease burdens. Innovations in nutrition science, including nanotechnology to enhance bioactive compound delivery, and systematic evaluation of whole-food, plant-based regimens, have highlighted the potential for sustainable, health-promoting dietary strategies. This chapter synthesizes evidence from key systematic reviews, cohort studies, and clinical trials to elucidate the impact of plant-based diets on nutrient status, chronic disease risk, and environmental sustainability.

4.1 Nutrient Intake and Status in Plant-Based Versus Omnivorous Diets

Neufingerl and Eilander (2022) systematically reviewed 141 studies comparing nutrient intake and biochemical status between adults adhering to vegetarian/vegan diets and omnivorous meat-eaters. They found that, although total energy intake did not significantly differ between groups, macronutrient distribution varied: protein intake averaged 12.9% and 13.4% of energy for vegans and vegetarians respectively versus 16.0% in meat-eaters, yet remained above the lower limit of the Acceptable Macronutrient Distribution Range (Neufingerl & Eilander, 2022). Fiber intake was significantly higher among vegans (mean = 44 g/day) and vegetarians (28 g/day) than among meat-eaters (21 g/day), with 74% of vegan cohorts meeting the Adequate Intake for fiber versus 6% of meat-eaters (Neufingerl & Eilander, 2022). Intake of polyunsaturated fatty acids (PUFAs) followed a vegan gradient 8.84% of energy, vegetarians 6.79%, and meat-eaters 5.95%, reflecting greater consumption of PUFA-rich plant oils and nuts (Neufingerl & Eilander, 2022). However, plant-based populations exhibited lower intakes and status of certain micronutrients. Vitamin B_{12} status was notably diminished in vegans (mean serum B_{12} = 122 pmol/L) versus meat-eaters (281 pmol/L), with 52% of vegans classified as deficient (<118 pmol/L). Calcium intake averaged 848 mg/day in vegans versus ~1100 mg/day in meat-eaters and vegetarians, potentially contributing to observed reductions in bone mineral density. Iodine status was also lower among vegans and some vegetarians, correlating with reduced dairy and seafood consumption. Long-chain omega-3 fatty acids (EPA and DHA) were markedly reduced in plant-based groups (e.g., 27 mg/day EPA and 4 mg/day DHA for vegans) compared with meat-eaters (94 mg/day and 172 mg/day respectively), despite higher α-linolenic acid (ALA) intake among vegans (2.01 g/day versus 1.38 g/day in meat-eaters) and the limited conversion efficiency of ALA to EPA/DHA (Neufingerl & Eilander, 2022). This evidence underscores the notion that well-planned plant-based diets can achieve adequate macronutrient profiles, particularly protein and fiber, but may require targeted strategies (e.g., supplementation, fortified foods) to ensure sufficient intake of vitamin B_{12}, vitamin D, calcium, iodine, and long-chain n-3 fatty acids.

4.2 Cardiovascular Health

Plant-based dietary patterns have been consistently associated with lower CVD incidence and improved risk factors. Satija and Hu (2018) conducted a comprehensive review showing that indices measuring healthy plant-based food intake, rich in whole grains, fruits, vegetables, nuts, and legumes; were inversely correlated with coronary heart disease and stroke events. Individuals adhering to high-quality plant-based diets exhibited up to a 32% lower risk of incident CVD (Satija & Hu, 2018). Trautwein and McKay (2020) examined how specific plant-based diet components influence dyslipidemia management. They found that dietary fiber, phytosterols, and monounsaturated and polyunsaturated fats prevalent in plant foods lower low-density lipoprotein cholesterol (LDL-C) by downregulating cholesterol synthesis and enhancing fecal bile-acid excretion. In clinical interventions, diets replacing saturated animal fats with plant-derived fats yielded significant reductions in total cholesterol (−13.5 mg/dL) and LDL-C (−9.2 mg/dL), contributing to a 20–25% decreased atherosclerotic risk (Trautwein & McKay, 2020). Salehin et al. (2023) reviewed randomized and observational studies, confirming that plant-based interventions reduce blood pressure, improve insulin sensitivity, and favorably modify lipid profiles, culminating in lower CVD morbidity and mortality (Salehin et al., 2023).

4.3 Type 2 Diabetes Prevention and Management

Type 2 diabetes mellitus (T2DM) prevention and treatment represent critical areas where plant-based diets excel. McMacken and Shah (2017) reviewed epidemiological cohorts and clinical trials demonstrating that vegan and vegetarian diets reduce T2DM prevalence and incidence. In the Adventist Health Study 2, vegans exhibited 49% lower odds of T2DM than nonvegetarians (OR = 0.51; 95% CI, 0.40–0.66), independent of body mass index (BMI) (McMacken & Shah, 2017). Prospective analyses showed that vegans had a 62% lower T2DM incidence over 2 years (OR = 0.38; 95% CI, 0.24–0.62) relative to meat-eaters, implying additional metabolic benefits beyond weight control. Randomized controlled trials offer further evidence. An intervention comparing a low-fat vegan diet (75% carbohydrates,

15% protein, 10% fat) with an American Diabetes Association-recommended omnivorous diet over 22 weeks reported a mean HbA_1c reduction of 1.23 percentage points in the vegan group versus 0.38 percentage points in controls ($p = 0.01$). Weight loss was also greater among vegans (−6.5 kg vs. −3.1 kg; $p < 0.001$); LDL-C decreased by 21.2% in the vegan cohort compared with 10.7% ($p = 0.02$). At 74 weeks, glycemia and lipid improvements persisted: vegans sustained an HbA_1c reduction of 0.40% compared with a 0.01% increase in controls (McMacken & Shah, 2017). These data illustrate that high-fiber, plant-based regimens directly enhance insulin sensitivity via improved postprandial glycemia, weight loss, and anti-inflammatory phytochemical effects.

4.4 Long-Term Health Outcomes (EPIC-Oxford)

The EPIC-Oxford cohort, as analyzed by Key, Papier, and Tong (2022), offers valuable longitudinal insights. Over ~20 years, vegetarians exhibited a 23% lower risk of incident ischemic heart disease (IHD) versus meat-eaters (HR = 0.77; 95% CI, 0.66–0.89), decreased to HR = 0.83 (95% CI, 0.71–0.98) after BMI adjustment. Vegans showed an 18% lower IHD risk (HR = 0.82; 95% CI, 0.60–1.13), although limited sample size widened the confidence intervals (Key et al., 2021). Lower LDL-C (−0.68 mmol/L in male vegans versus meat-eaters) and reduced systolic blood pressure (−2.6 mmHg in male vegans versus meat-eaters) partly mediated these cardiovascular benefits (Key et al., 2021). Regarding T2DM, vegetarians had 26% lower incidence (HR = 0.74; 95% CI, 0.66–0.84), and vegans had 34% lower risk (HR = 0.66; 95% CI, 0.48–0.92) than meat-eaters; adjustment for BMI attenuated these associations yet remained significant for vegetarians (HR = 0.86; 95% CI, 0.76–0.97) (Key et al., 2021). Other favorable outcomes included lower risks of diverticular disease, kidney stones, and cataracts among vegetarians, with vegans demonstrating lower cataract incidence. Conversely, vegetarian and vegan participants exhibited lower heel bone mineral density (t-scores: −0.80 for vegan women versus −0.54 for meat-eating women) and higher fracture risk, reflecting potential calcium and vitamin D insufficiencies (Key et al., 2021).

4.5 Cancer Prevention and Prognosis

The relationship between plant-based diets and cancer risk/prognosis has been explored in multiple meta-analyses and reviews. Zhao et al. (2022) conducted a meta-analysis of 3,059,009 subjects, finding that greater adherence to plant-based diets was associated with significantly reduced risks of colorectal (RR = 0.85; 95% CI, 0.80–0.91), pancreatic (RR = 0.74; 95% CI, 0.61–0.91), and gastric cancers (RR = 0.61; 95% CI, 0.42–0.89) (Zhao et al., 2022). Likewise in the same year Hardt et al., reviewed cancer prognosis, concluding that postdiagnosis consumption of whole grains, fruits, and vegetables improved overall survival and disease-free survival among breast and colorectal cancer survivors, likely mediated by anti-inflammatory and antioxidant phytochemicals (Hardt et al., 2022).

4.6 Obesity, Weight Management, and Chronic Kidney Disease

Ivanova et al. (2021) reviewed mechanisms by which plant-based diets promote weight control. Higher dietary fiber intake enhances satiety and stabilizes glycemia, whereas lower dietary energy density fosters a negative energy balance. Moreover, plant-based regimens increase gut microbiome diversity, favoring short-chain fatty-acid production, which modulates adiposity and insulin sensitivity (Ivanova et al., 2021). In metabolic ward studies, high-fiber, high-carbohydrate diets (70% carbohydrates, <10% fat, 65 g fiber/day) reduced insulin requirements from 26 to 11 units/day in T2DM patients, demonstrating intrinsic insulin-sensitizing properties beyond weight loss (McMacken & Shah, 2017). For chronic kidney disease (CKD), several authors emphasized that plant-dominant diets slow CKD progression by reducing acid load, decreasing phosphorus bioavailability, and attenuating uremic toxin generation. High-fiber intake improves gut-barrier function, reducing systemic inflammation and endotoxemia, which in turn lowers cardiovascular risks, a critical benefit given the elevated CVD mortality among CKD patients (Joshi et al., 2021).

4.7 Environmental Sustainability

Plant-based diets align with planetary health by significantly reducing the environmental footprint. Craig et al. (2021) quantified that tofu-derived protein requires 74 times less land, 8 times less water, and emits 25 times fewer greenhouse gases than equivalent beef protein (Craig et al., 2021). Neufingerl and Eilander (2022) reported that widespread adoption of plant-based patterns could decrease dietary greenhouse-gas emissions by 54–87%, land use by ~42%, and freshwater use by ~28%. Furthermore, McMacken and Shah (2017) noted that reduced dependence on animal agriculture curtails pharmaceutical pollutants through decreased medication requirements, as well as diminishing agricultural runoff and biodiversity loss (McMacken & Shah, 2017; Neufingerl & Eilander, 2022).

4.8 Challenges and Future Directions

Robust evidence from observational cohorts, meta-analyses, and clinical trials converges on the conclusion that well-designed plant-based diets emphasizing whole grains, legumes, vegetables, fruits, nuts, and seeds substantially reduce the risk of

CVD, type 2 diabetes, several cancers, obesity, and CKD, while also lowering all-cause mortality (Key et al., 2021; McMacken & Shah, 2017; Neufingerl & Eilander, 2022; Satija & Hu, 2018; Tan et al., 2024). Mechanisms include improvements in lipid profiles, enhanced insulin sensitivity, reduced inflammation, favorable gut microbiome modulation, and decreased exposure to dietary saturated fats and heme iron. Potential nutrient insufficiencies, particularly vitamin B_{12}, vitamin D, calcium, iodine, and long-chain omega-3 fatty acids, must be addressed through fortified foods or supplementation. From an ecological standpoint, plant-based dietary transitions offer a viable strategy to mitigate greenhouse-gas emissions, land and water use, pharmaceutical pollution, and biodiversity threats (Craig et al., 2021; Neufingerl & Eilander, 2022). Public health and clinical guidelines should therefore prioritize plant-forward nutrition, coupled with educational initiatives and policy support, to realize dual benefits for human and planetary health.

5 Traditional Diets as Sustainable Alternatives: Focus on the Milpa Diet

Traditional dietary patterns are increasingly recognized for their potential to improve public health and environmental sustainability. In Mexico and Mesoamerica, the Milpa Diet, rooted in the intercropping of maize (*Zea mays*), common bean (*Phaseolus vulgaris* and other *Phaseolus species*.), squash (*Cucurbita* spp.), chili varieties (*Capsicum* spp.), and other locally adapted species such as tomato, quelites, amaranth, chia, among others, represents a model of nutritional complementarity and ecological balance (Ballhorn et al., 2012; Valerino-Perea et al., 2019). This review explores the evidence supporting the role of the Milpa Diet in preventing NCDs and promoting environmental stewardship.

5.1 Composition and Historical Relevance of the Milpa Diet

The Milpa system originated in the Archaic period of Mesoamerica, where wild progenitors of maize, beans, and squashes were managed in a single agroecosystem, facilitating nutrient complementarity: maize providing carbohydrates, beans supplying protein and micronutrients, and squash offering vitamins and fiber (Ballhorn et al., 2012). Subsequent anthropological studies have documented that this triad, along with chili, tomato, and amaranth, has persisted as the backbone of rural diets (Valerino-Perea et al., 2019). Modern definitions of the Milpa Diet emphasize these core elements, plus seasonal fruits, edible herbs, and minor animal products, creating a culturally concordant, plant-based dietary pattern (Biruete et al., 2024; Hernández Galindo et al., 2022).

5.2 *Impact on Noncommunicable Chronic Diseases*

5.2.1 Metabolic Health and Obesity

Dietary patterns rich in complex carbohydrates, plant proteins, and dietary fiber hallmarks of the Milpa Diet are associated with improved weight control and insulin sensitivity. Experimental studies with foods of Mesoamerican origin have demonstrated their bioactive effects on adipose and liver metabolism, including modulation of lipid accumulation and enhancement of antioxidant defenses (Meza-Rios et al., 2023). In cross-sectional surveys, higher consumption of traditional maize–bean–squash combinations correlated with lower body mass index and reduced markers of metabolic syndrome compared with Westernized diets high in ultra-processed foods (Moreno-Altamirano et al., 2017; Oviedo-Solís et al., 2022).

5.2.2 Cardiovascular Disease and Dyslipidemia

The synergy of fiber-rich legumes and monounsaturated fatty acids from squash seeds and squash oil contributes to improved lipid profiles. In clinical cohorts, adherence to plant-dominant, culturally congruent diets such as the Milpa pattern yielded significant reductions in low-density lipoprotein cholesterol and inflammatory markers (Biruete et al., 2024; Trautwein & McKay, 2020). Although specific trials on the Milpa Diet are limited, analogous data from plant-based Mexican dietary interventions support its cardioprotective potential.

5.2.3 Type 2 Diabetes Mellitus

High-fiber, low-glycemic-index staples such as nixtamalized maize and common bean reduce postprandial glucose excursions and improve glycemic control in individuals with prediabetes and type 2 diabetes (McMacken & Shah, 2017; Valerino-Perea et al., 2019). Preliminary interventional studies integrating Milpa-based meal plans have shown decreases in glycosylate hemoglobin (HbA_1c) of up to 1 percentage point over 12 weeks, outperforming standard dietary advice (Huerta-Álvarez et al., 2025).

5.2.4 Chronic Kidney Disease

The plant-protein dominance of the Milpa Diet offers moderate phosphorus bioavailability and a favorable acid–base balance, which are critical to CKD management. A narrative review highlighted that Hispanic patients following Milpa-like patterns experienced stabilized glomerular filtration rates and reduced progression to end-stage renal disease, attributed to lower dietary acid load and enhanced gut-microbiota-mediated anti-inflammatory effects (Biruete et al., 2024).

5.2.5 Contribution to Environmental Sustainability

The Milpa agroecosystem exemplifies biodiversity conservation and resource efficiency. Intercropping reduces land use by up to 50% compared with monocultures, enhances soil fertility through nitrogen fixation by beans, and lowers greenhouse-gas emissions by integrating perennial herbs and cover crops (Hernández Galindo et al., 2022). Life-cycle assessments indicate that a shift from animal-based staples to Milpa-Diet components can reduce the dietary carbon footprint by approximately 30% while maintaining nutrient adequacy (Gémez et al., 2019).

5.2.6 Challenges and Future Directions

Despite compelling evidence, the Milpa Diet faces challenges of urbanization, market globalization, and dietary acculturation that favor ultra-processed foods (Oviedo-Solís et al., 2022). Research gaps include long-term randomized trials quantifying disease endpoints, optimization of portion sizes for clinical subgroups, and policy frameworks that incentivize Milpa-based food systems (Meza-Rios et al., 2023; Valerino-Perea et al., 2019).

The Milpa Diet, anchored in maize, beans, squash, and complementary local species, offers a sustainable, culturally resonant strategy for mitigating NCDs and reducing environmental impact. Its nutrient synergy, aligned with modern nutritional science and agroecological principles, warrants investment in clinical research, public-health promotion, and agricultural policies to revitalize this ancestral dietary model.

6 Precision Nutrition and Nutritional Genomics in Chronic Disease Management

6.1 Obesity

A growing field of nutrigenomic has evidenced a complex network of gene–diet interactions underlying chronic-disease susceptibility, revealing not only metabolic but also sensory and behavioral dimensions to nutritional responses. For instance, the *TAS1R2* gene, encoding a sweet taste receptor, harbors the rs35874116 variant, which has been associated with higher carbohydrate intake and preference for sweetened foods, factors directly contributing to glycemic dysregulation and obesity risk (Ponnusamy et al., 2024). Likewise, polymorphisms in *CD36*, particularly rs1761667, influence fat-taste sensitivity and lipid-absorption efficiency, thereby affecting dietary fat preference (Jaime-Lara et al., 2023). The *FABP2* rs1799883 variant alters intestinal fatty-acid transport and has been linked to elevated postprandial triglyceride levels and increased cardiometabolic risk. Beyond metabolism, *DRD2* rs1800497, which modulates dopaminergic signaling along the gut–brain axis, has been implicated in emotional eating and dysbiosis-associated obesity phenotypes (Obregón et al., 2022). In contrast, higher copy-number variation of the *AMY1* gene coding for salivary amylase has been correlated with improved starch digestion and a lower risk of central adiposity, suggesting a protective genotype–diet alignment in high-carbohydrate traditional diets (Venkatapoorna et al., 2019). These variants often interact with dietary patterns such as high-fat or high-glycemic-load diets, exemplify the polygenic architecture of nutritional response, and provide a molecular basis for tailoring interventions in the management and prevention of obesity, type 2 diabetes, and CVD. The integration of these insights into clinical frameworks enhances our capacity to stratify individuals, not only by metabolic phenotype but also by sensory, behavioral, and digestive genotypes, supporting a truly multidimensional model of precision nutrition (Mansour et al., 2024).

6.2 Type 2 Diabetes-Related Genes

In this context, multiple genetic variants have been identified that modulate insulin sensitivity, beta-cell function, and incretin response, all of which shape dietary efficacy. The *TCF7L2* rs7903146 variant is among the most consistently replicated polymorphisms associated with impaired insulin secretion and increased diabetes risk (Ding et al., 2018). Carriers of the T allele exhibit enhanced glycemic control when following a Mediterranean-style diet, rich in monounsaturated fats and complex carbohydrates, suggesting a diet–genotype synergy that can mitigate genetic risk (Barabash et al., 2020). Additionally, polymorphisms in *IRS1* (rs2943641) and *GIPR* (rs10423928) affect insulin signaling and incretin hormone activity respectively, and have been shown to influence postprandial glucose levels depending on dietary carbohydrate quality (Brown et al., 2023). These insights support a stratified nutritional approach, where individuals with impaired insulin signaling pathways might benefit more from low-glycemic or higher-fiber dietary patterns.

6.3 Cardiovascular Health

Regarding CVD, nutrigenomic strategies highlight the role of *APOE*, *ACE*, and *ADRB2* gene variants in modulating lipid metabolism, blood pressure, and adrenergic regulation. The *APOE* ε4 allele, for instance, predisposes carriers to elevated LDL-C levels and poorer response to saturated fat, but shows improved lipid profiles under diets enriched with omega-3 fatty acids and plant sterols (Abumweis et al., 2023; Ismail et al., 2024). Meanwhile, the *ACE* I/D polymorphism, influencing angiotensin-converting enzyme activity, has been linked to differential blood pressure response to sodium intake, with D allele carriers exhibiting more pronounced reductions under sodium-restricted or Dietary Approaches to Stop Hypertension-like diets (Maris et al., 2019). Moreover, *ADRB2* rs1042713 (Arg16Gly) affects vascular tone and metabolic rate, and its interaction with dietary fat has been associated with variation in hypertension and lipid responses (de Luis et al., 2023). These findings underscore the necessity of tailoring cardiovascular prevention strategies, not only around dietary patterns but also around individual genetic architecture.

6.4 Precision Nutrition, Metabolic Management Challenges, and Future Directions

Several studies have investigated the utility of genetic architecture in guiding nutritional therapies, among which the POUNDS Lost trial stands out as one of the most comprehensive. This 2-year randomized controlled trial enrolled over 800 overweight adults, assigning them to one of four dietary patterns varying in macronutrient composition (Bray et al., 2024). Building on its robust design and long-term follow-up, recent analyses have expanded our understanding of how genetic variants, epigenetic markers, and metabolomic profiles can influence individual responses to dietary interventions (Qi et al., 2023).

Notable findings include evidence that *FGF21* gene variants modulate the effect of carbohydrate-focused diets on central adiposity, participants with specific *FGF21* alleles experienced significantly greater reductions in trunk fat when following low-carbohydrate diets, and the inverse was also observed (De Sousa-Coelho et al., 2023; Heianza et al., 2016). In parallel, the *IRS1* rs2943641 variant has consistently shown clinical relevance, with individuals carrying the CC genotype achieving superior metabolic improvements such as enhanced insulin sensitivity and reduced waist circumference when following high-carbohydrate, low-fat diets (Fajardo et al., 2023; Huang et al., 2016). Furthermore, recent analyses involving genetic risk scores (GRS), which aggregate multiple single-nucleotide polymorphisms (SNPs) including *AMY1* copy number variants, suggest that individuals with specific GRS profiles exhibit greater improvements in body weight and waist circumference when adhering to low-carbohydrate regimens (Bray et al., 2024).

These findings collectively underscore the growing potential of genotype-informed dietary prescriptions in enhancing the effectiveness of weight-loss strategies. Although the effect sizes remain moderate, and further adjustments are made by ethnicity, the consistent gene–diet interactions observed support a precision nutrition approach that goes beyond phenotype alone, incorporating genetic and molecular data to optimize clinical outcomes.

6.5 Final Remarks

As the burden of noncommunicable diseases continues to rise in tandem with escalating environmental and social inequities, dietary therapy emerges not merely as a clinical intervention but as a multidimensional strategy at the intersection of health, sustainability, and justice. From precision nutrition guided by genetic insights to the cultural resilience of traditional diets such as the Milpa, the evidence underscores the potential of dietary interventions to transform individual and population health outcomes while safeguarding ecological balance.

Moving forward, researchers and clinicians must embrace integrative approaches that consider sociocultural identity, genetic predisposition, environmental impact, and mental health as interconnected pillars of dietary health. Policymakers are also called to act: supporting systems that democratize access to nutritious, culturally resonant, and environmentally sustainable foods is not an adjunct goal—it is a public health imperative. Dietary therapy, when informed by both ancient wisdom and cutting-edge science, holds the power to not only reduce disease burden but also anchor a collective transition toward planetary health. Aligning our plates with global goals is no longer aspirational—it is essential.

References

Abumweis, S., Alzyoud, L., & Alqadi, S. (2023). Apolipoprotein E genetic variant and blood lipid responses to plant sterols: A systematic review and pooled analysis of clinical trials. *Preventive Nutrition and Food Science, 28*(4), 377–385. https://doi.org/10.3746/PNF.2023.28.4.377

Ballhorn, D. J., Kautz, S., Heil, M., & Hegeman, A. D. (2012). The archaic diet in Mesoamerica: Incentive for Milpa development and species domestication. *Economic Botany, 66*(4), 328–343., 4(5). https://doi.org/10.1371/JOURNAL.PONE.0005450

Barabash, A., Valerio, J. D., Garcia de la Torre, N., Jimenez, I., del Valle, L., Melero, V., Assaf-Balut, C., Fuentes, M., Bordiu, E., Durán, A., Herraiz, M. A., Izquierdo, N., Torrejón, M. J., de Miguel, P., Runkle, I., Rubio, M. A., & Calle-Pascual, A. L. (2020). TCF7L2 rs7903146 polymorphism modulates the association between adherence to a Mediterranean diet and the risk of gestational diabetes mellitus. *Metabolism Open, 8*, 100069. https://doi.org/10.1016/j.metop.2020.100069

Biruete, A., Leal-Escobar, G., Espinosa-Cuevas, Á., Mojica, L., & Kistler, B. M. (2024). Dieta de la Milpa: A culturally-concordant plant-based dietary pattern for Hispanic/Latine people with chronic kidney disease. *Nutrients, 16*(5). https://doi.org/10.3390/NU16050574

Bray, G. A., Qi, L., & Sacks, F. M. (2024). Is there an ideal diet? Some insights from the POUNDS lost study. *Nutrients, 16*(14). https://doi.org/10.3390/NU16142358

Brown, J. E., Pham, T., Burden, H., & Braakhuis, A. J. (2023). Specific genotypes associated with differences in fasting insulin levels and body mass index in healthy young males: Implications for gene–nutrient interactions – An exploratory study. *Current Developments in Nutrition, 7*(11). https://doi.org/10.1016/j.cdnut.2023.102018

Craig, W. J., Mangels, A. R., Fresán, U., Marsh, K., Miles, F. L., Saunders, A. V., Haddad, E. H., Heskey, C. E., Johnston, P., Larson-Meyer, E., & Orlich, M. (2021). The safe and effective use of plant-based diets with guidelines for health professionals. *Nutrients, 13*(11). https://doi.org/10.3390/NU13114144

De Luis, D. A., Izaola, O., Primo, D., & Gómez, J. J. L. (2023). Role of beta-2 adrenergic receptor polymorphism (rs1042714) on body weight and glucose metabolism response to a meal-replacement hypocaloric diet. *Nutrition, 116*. https://doi.org/10.1016/j.nut.2023.112170

De Sousa-Coelho, A. L., Rodriguez-Rodriguez, R., Softic, S., Jonker, J. W., & Relat, J. (2023). Editorial: FGF21 as a therapeutic target for obesity and insulin resistance: From rodent models to humans. *Frontiers in Endocrinology, 14*. https://doi.org/10.3389/FENDO.2023.1253675

Ding, W., Xu, L., Zhang, L., Han, Z., Jiang, Q., Wang, Z., & Jin, S. (2018). Meta-analysis of association between TCF7L2 polymorphism rs7903146 and type 2 diabetes mellitus. *BMC Medical Genetics, 19*(1). https://doi.org/10.1186/S12881-018-0553-5

Fajardo, C. M., Cerda, A., Bortolin, R. H., de Oliveira, R., Stefani, T. I. M., dos Santos, M. A., Braga, A. A., Dorea, E. L., Bernik, M. M. S., Bastos, G. M., Sampaio, M. F., Damasceno, N. R. T., Verlengia, R., de Oliveira, M. R. M., Hirata, M. H., & Hirata, R. D. C. (2023). Influence of polymorphisms in IRS1, IRS2, MC3R, and MC4R on metabolic and inflammatory status and food intake in Brazilian adults: An exploratory pilot study. *Nutrition Research, 119*, 21–32. https://doi.org/10.1016/j.nutres.2023.08.008

Gémez, G., Fisberg, R. M., Previdelli, Á. N., Sales, C. H., Kovalskys, I., Fisberg, M., Herrera-Cuenca, M., Sanabria, L. Y. C., García, M. C. Y., Torres, R. G. P., Rigotti, A., Guajardo, V., Zimberg, I. Z., Chinnock, A., Murillo, A. G., & Brenes, J. C. (2019). Diet quality and diet diversity in eight Latin American countries: Results from the Latin American study of nutrition and health (ELANS). *Nutrients, 11*(7). https://doi.org/10.3390/NU11071605

Hardt, L., Mahamat-Saleh, Y., Aune, D., & Schlesinger, S. (2022). Plant-based diets and cancer prognosis: A review of recent research. *Current Nutrition Reports, 11*(4), 695–716. https://doi.org/10.1007/S13668-022-00440-1

Heianza, Y., Ma, W., Huang, T., Wang, T., Zheng, Y., Smith, S. R., Bray, G. A., Sacks, F. M., & Qi, L. (2016). Macronutrient intake-associated FGF21 genotype modifies effects of weight-loss diets on 2-year changes of central adiposity and body composition: The POUNDS lost trial. *Diabetes Care, 39*(11), 1909–1914. https://doi.org/10.2337/DC16-1111

Hernández-Galindo, H. S., Alanís-García, E., & Omaña-Covarrubias, A. (2022). La Dieta de La Milpa: Como una alternativa en salud pública en el Valle del Mezquital Hidalguense, después de la pandemia de la covid-19: The Milpa diet: As an alternative in public health in the valley of the Mezquital, state of Hidalgo, after the covid-19 pandemic. *Educación y Salud Boletín Científico Instituto de Ciencias de La Salud Universidad Autónoma Del Estado de Hidalgo, 10*(20), 7–20. https://doi.org/10.29057/ICSA.V10I20.8362

Huang, T., Ley, S. H., Zheng, Y., Wang, T., Bray, G. A., Sacks, F. M., & Qi, L. (2016). Genetic susceptibility to diabetes and long-term improvement of insulin resistance and β cell function during weight loss: The preventing overweight using novel dietary strategies (POUNDS LOST) trial. *American Journal of Clinical Nutrition, 104*(1), 198–204. https://doi.org/10.3945/ajcn.115.121186

Huerta-Álvarez, A., Arellano, M., Chávez-Méndez, C. A., Carpinteyro-Espin, P., Palacios-Reyes, C., & Pérez-Escobar, J. (2025). Milpa diet for MASLD in Mesoamerican populations: Feasibility, advantages, and future perspectives. *Life (Basel, Switzerland), 15*(5), 812. https://doi.org/10.3390/LIFE15050812

Ismail, A. B., Balcıoğlu, Ö., Özcem, B., & Ergoren, M. Ç. (2024). APOE gene variation's impact on cardiovascular health: A case-control study. *Biomedicine, 12*(3). https://doi.org/10.3390/BIOMEDICINES12030695

Ivanova, S., Delattre, C., Karcheva-Bahchevanska, D., Benbasat, N., Nalbantova, V., & Ivanov, K. (2021). Plant-based diet as a strategy for weight control. *Food, 10*(12). https://doi.org/10.3390/FOODS10123052

Jaime-Lara, R. B., Brooks, B. E., Vizioli, C., Chiles, M., Nawal, N., Ortiz-Figueroa, R. S. E., Livinski, A. A., Agarwal, K., Colina-Prisco, C., Iannarino, N., Hilmi, A., Tejeda, H. A., & Joseph, P. V. (2023). A systematic review of the biological mediators of fat taste and smell. *Physiological Reviews, 103*(1), 855–918. https://doi.org/10.1152/PHYSREV.00061.2021

Joshi, S., McMacken, M., & Kalantar-Zadeh, K. (2021). Plant-based diets for kidney disease: A guide for clinicians. *American Journal of Kidney Diseases, 77*(2), 287–296. https://doi.org/10.1053/j.ajkd.2020.10.003

Key, T. J., Papier, K., & Tong, T. Y. N. (2021). Plant-based diets and long-term health: Findings from the EPIC-Oxford study. *Proceedings of the Nutrition Society, 81*(2). https://doi.org/10.1017/S0029665121003748

Kim, M. T., Heitkemper, E. M., Hébert, E. T., Hecht, J., Crawford, A., Nnaka, T., Hutson, T. S., Rhee, H., & Radhakrishnan, K. (2022). Redesigning culturally tailored intervention in the precision health era: Self-management science context. *Nursing Outlook, 70*(5), 710–724. https://doi.org/10.1016/J.OUTLOOK.2022.05.015

Klapp, A. L., Feil, N., & Risius, A. (2022). A global analysis of National Dietary Guidelines on plant-based diets and substitutions for animal-based foods. *Current Developments in Nutrition, 6*(11). https://doi.org/10.1093/cdn/nzac144

Lane, M. M., Gamage, E., Travica, N., Dissanayaka, T., Ashtree, D. N., Gauci, S., Lotfaliany, M., O'Neil, A., Jacka, F. N., & Marx, W. (2022). Ultra-processed food consumption and mental health: A systematic review and meta-analysis of observational studies. *Nutrients, 14*(13). https://doi.org/10.3390/NU14132568

Loughman, A., Staudacher, H. M., Rocks, T., Ruusunen, A., Marx, W., O'Neil, A., & Jacka, F. N. (2021). Diet and mental health. *Modern Trends in Psychiatry, 32*, 100–112. https://doi.org/10.1159/000510422

Mansour, S., Alkhaaldi, S. M. I., Sammanasunathan, A. F., Ibrahim, S., Farhat, J., & Al-Omari, B. (2024). Precision nutrition unveiled: Gene–nutrient interactions, microbiota dynamics, and lifestyle factors in obesity management. *Nutrients, 16*(5). https://doi.org/10.3390/NU16050581

Maris, S. A., Williams, J. S., Sun, B., Brown, S., Mitchell, G. F., & Conlin, P. R. (2019). Interactions of the DASH diet with the renin-angiotensin-aldosterone system. *Current Developments in Nutrition, 3*(9). https://doi.org/10.1093/cdn/nzz091

McMacken, M., & Shah, S. (2017). A plant-based diet for the prevention and treatment of type 2 diabetes. *Journal of Geriatric Cardiology, 14*(5), 342–354. https://doi.org/10.11909/J.ISSN.1671-5411.2017.05.009

Meza-Rios, A., López-Villalobos, E. F., Anguiano-Sevilla, L. A., Ruiz-Quezada, S. L., Velazquez-Juarez, G., López-Roa, R. I., Marin-Molina, A. L., & Zepeda-Morales, A. S. M. (2023). Effects of foods of Mesoamerican origin in adipose tissue and liver-related metabolism. *Medicina (Lithuania), 59*(11). https://doi.org/10.3390/MEDICINA59111907

Moreno-Altamirano, L., García-García, J. J., Salvatore, P., Soto-Estrada, G., & Hernández-Montoya, D. (2017). Metabolic syndrome: Changes in mediterranean and Mesoamerican diet due to socioeconomic factors in Mexico and Italy. *Mediterranean Journal of Nutrition and Metabolism, 10*(1), 49–59. https://doi.org/10.3233/MNM-16124;SUBPAGE:STRING:FULL

Neufingerl, N., & Eilander, A. (2022). Nutrient intake and status in adults consuming plant-based diets compared to meat-eaters: A systematic review. *Nutrients, 14*(1). https://doi.org/10.3390/NU14010029

Obregón, A. M., Oyarce, K., García-Robles, M. A., Valladares, M., Pettinelli, P., & Goldfield, G. S. (2022). Association of the dopamine D2 receptor rs1800497 polymorphism with food addiction, food reinforcement, and eating behavior in Chilean adults. *Eating and Weight Disorders, 27*(1), 215–224. https://doi.org/10.1007/S40519-021-01136-1

Oviedo-Solís, C. I., Monterrubio-Flores, E. A., Cediel, G., Denova-Gutiérrez, E., & Barquera, S. (2022). Trend of ultraprocessed product intake is associated with the double Burden of malnutrition in Mexican children and adolescents. *Nutrients, 14*(20). https://doi.org/10.3390/NU14204347

Ponnusamy, V., Subramanian, G., Vasanthakumar, K., Muthuswamy, K., Panneerselvan, P., Krishnan, V., & Subramaniam, S. (2024). T1R2/T1R3 polymorphism affects sweet and fat perception: Correlation between SNP and BMI in the context of obesity development. *Human Genetics, 144*(1). https://doi.org/10.1007/S00439-024-02690-1

Qi, L., Heianza, Y., Li, X., Sacks, F. M., & Bray, G. A. (2023). Toward precision weight-loss dietary interventions: Findings from the POUNDS lost trial. *Nutrients, 15*(16). https://doi.org/10.3390/NU15163665

Salehin, S., Rasmussen, P., Mai, S., Mushtaq, M., Agarwal, M., Hasan, S. M., Salehin, S., Raja, M., Gilani, S., & Khalife, W. I. (2023). Plant based diet and its effect on cardiovascular disease. *International Journal of Environmental Research and Public Health, 20*(4). https://doi.org/10.3390/IJERPH20043337

Satija, A., & Hu, F. B. (2018). Plant-based diets and cardiovascular health. *Trends in Cardiovascular Medicine, 28*(7), 437–441. https://doi.org/10.1016/j.tcm.2018.02.004

Tan, J., Zhang, S., Jiang, Y., Li, J., & Yang, C. (2024). Plant-based diet and risk of all-cause mortality: A systematic review and meta-analysis. *Frontiers in Nutrition, 11*. https://doi.org/10.3389/FNUT.2024.1481363

Trautwein, E. A., & McKay, S. (2020). The role of specific components of a plant-based diet in management of dyslipidemia and the impact on cardiovascular risk. *Nutrients, 12*(9), 1–21. https://doi.org/10.3390/NU12092671

Valerino-Perea, S., Lara-Castor, L., Armstrong, M. E. G., & Papadaki, A. (2019). Definition of the traditional Mexican diet and its role in health: A systematic review. *Nutrients, 11*(11). https://doi.org/10.3390/NU11112803

Venkatapoorna, C. M. K., Ayine, P., Parra, E. P., Koenigs, T., Phillips, M., Babu, J. R., Sandey, M., & Geetha, T. (2019). Association of Salivary Amylase (AMY1) gene copy number with obesity in Alabama elementary school children. *Nutrients, 11*(6). https://doi.org/10.3390/NU11061379

Vilar-Compte, M., Burrola-Méndez, S., Lozano-Marrufo, A., Ferré-Eguiluz, I., Flores, D., Gaitán-Rossi, P., Teruel, G., & Pérez-Escamilla, R. (2021). Urban poverty and nutrition challenges associated with accessibility to a healthy diet: A global systematic literature review. *International Journal for Equity in Health, 20*(1). https://doi.org/10.1186/S12939-020-01330-0

Zhao, Y., Zhan, J., Wang, Y., & Wang, D. (2022). The relationship between plant-based diet and risk of digestive system cancers: A meta-analysis based on 3,059,009 subjects. *Frontiers in Public Health, 10*. https://doi.org/10.3389/FPUBH.2022.892153/PDF

Chapter 6
Plants as a Friendly and Emerging Alternative for Disease Management in Humans

David Gilberto García-Hernández, Alexander Joseph Pawlukiewicz, and Edelia Claudina Villarreal-Ibarra

1 Introduction

Medicinal and aromatic plants, especially ethnopharmacologically, have been used since time immemorial as a source of medicines and health care (Chaachouay & Zidane, 2024). These popular medicines were mainly simple preparations such as tinctures, macerations, teas, infusions, decoctions, and other herbal preparations (Benkhnigue et al., 2022). The right dose of the plant and the mode of administration for certain diseases were transmitted by oral tradition from one generation to the other and the traditional pharmacopeias noted the knowledge of these medicinal plants (Avery & Hains, 2017; d'Avigdor et al., 2014; Balunas & Kinghorn, 2005). Researchers usually gather and identify plants of interest (Balunas & Kinghorn, 2005). The use of plants has been helpful to scientists in the discovery of some of the first drugs to have unique properties. As an example, Silymarin, an extract from the seeds of *Silybum marianum*, has been used to treat hepatitis. Digitoxin, which is obtained from the foxglove plant, has been used to treat heart disease, specifically congestive heart failure (Kumar Shakya & Arvind Kumar Shakya, 2016).

D. G. García-Hernández (✉)
Facultad de Ciencias Biológicas, Departamento de Química, Universidad Autónoma de Nuevo León, San Nicolás de los Garza, Mexico
e-mail: david.garciahrz@uanl.edu.mx

A. J. Pawlukiewicz
Department of Integrative Biology, Ostling Laboratory, University of Texas at Austin, Austin, TX, USA
e-mail: alexander.pawlukiewicz@austin.utexas.edu

E. C. Villarreal-Ibarra
División de Ciencias Básicas e Ingeniería, Universidad Popular de la Chontalpa, Cárdenas, Mexico
e-mail: edelia.villareal@upch.mx

M. S. Heya (ed.), *Natural Alternatives and Nanotechnology Applied to the One Health Approach*, https://doi.org/10.1007/978-3-032-03675-9_6

In summary, plants are a source of natural products known as secondary metabolites, which exhibit a wide range of biological activities that impact human daily life. Their beneficial role extends beyond nutrition to include applications in human health, making them a valuable resource. However, the search for these metabolites must employ specific methodologies to prevent the indiscriminate use of plant species (which risks driving them to extinction) and to minimize the generation of environmentally harmful byproducts. Such efforts should prioritize interdisciplinary collaboration to reduce uncertainty in result interpretation and foster sustainable scientific progress.

2 Plants as a Friendly and Emerging Alternative for Disease Management in Humans

Among the primordial organisms of planet Earth, we have the Plantae kingdom, a highly diverse group on planet Earth, whose origins are traced back to a clade called "charophytes" approximately 476–432 million years ago (Arana et al., 2014).

The use of them by humanity is evidenced by fossil and archeological sources, whether in paintings, writing, or via oral transmission across generations. Over time, the benefits and applications derived from them thanks to the development of new technologies and the application of targeted knowledge have expanded from basic sustenance, clothing, shelter, and disease treatment to broader uses.

The use of herbs for medicinal purposes dates back to antiquity, with archeological evidence tracing back to the Paleolithic era, 60,000 years ago. The Sumerians documented herbal treatments over 5000 years ago, whereas the Egyptians, Greeks, and Romans also employed them, although there were fluctuations in their popularity. These historical traditions have significantly influenced the development of modern pharmacology and medicine, demonstrating a longstanding relationship between humanity and plant-based remedies (Chaachouay & Zidane, 2024).

Traditional Chinese Medicine practitioners use herbal and plant-based formulas to promote health and stimulate organ function, with their role extending beyond the chemical composition and intrinsic properties of the plants. That is, the components of these formulas, some of which have been used in China for 2200 years combine their functions in a complementary manner, unlike Western medicine. Traditional Chinese Medicine emphasizes the synergy between herbs to enhance their benefits, reflecting a treatment philosophy rooted in holistic complementarity (Fan et al., 2012; He, 2013).

Traditional pharmacopeias, underpinned by the wisdom and methods of native societies, often rely on plant-based materials containing secondary metabolites. Herbal preparations and remedies stemming from traditional knowledge were passed on for several centuries and still occupy important positions in health care practices from various cultures. These formulations are considered by many as beyond medicinal but rather cultural (Chaachouay & Zidane, 2024).

Primary metabolites such as lipids, carbohydrates, and amino acids are produced in large amounts and are essential to plant structure, whereas secondary metabolites produced in smaller quantities have a physiological role and help the plant to combat various kinds of stresses, deficiency of nutrients, and pathogen attack. These secondary metabolites, which include alkaloids, terpenoids, phenolics, and glycosides, exhibit antimicrobial activity; they have protective functions against pathogenic attacks and are involved in communication between plant cells. The secondary metabolites are generally divided into four major groups, namely terpenes, phenolics, and nitrogen- and sulfur-containing compounds (Jalota et al., 2024).

Rodrigues (2025) emphasizes the significant role of natural products derived from plants in advancing medicine, cosmetics, agriculture, and sustainability. Key examples include paclitaxel from the Pacific yew tree (chemotherapy), artemisinin from sweet wormwood (malaria treatment), shea butter and aloe vera in skincare, and neem oil (a natural pesticide). Plant-derived antioxidants such as polyphenols (green tea) and retinoids (papaya) are vital in anti-aging cosmetics. These products showcase nature's chemical diversity and bioactivity, offering sustainable alternatives to synthetic solutions. Their terrestrial origins underscore nature's potential as a continued source of innovation for improving human health and ecological balance.

Throughout history, natural compounds have played a key role in drug development, particularly for treating infectious diseases (Atanasov et al., 2021).

Many of these substances have long been used in traditional medicine and are now being studied for their medicinal properties. The discovery and analysis of these compounds have driven the development of new drugs and treatments for a wide variety of diseases and conditions. These substances exhibit distinct chemical structures and diverse pharmacological effects. The chemical diversity of plant-derived natural compounds makes them a valuable resource for identifying novel drug candidates. Pharmaceutical researchers can explore this vast repository of molecules to uncover new therapeutic agents for clinical applications(Chaachouay and Zidane, 2024).

Nowadays, the biological activity of many plants is studied and understood, along with the isolation and elucidation of the active compound(s) that they contain. The study begins by identifying the problem in question, followed by consulting traditional uses and customs. Once the data are gathered, field collection is carried out or sourced through a commercial chain that ensures proper plant identification.

The plant in question is washed and dried using drying ovens at temperatures no higher than 40 °C to avoid degrading thermolabile compounds, such as polyphenols. Once dried, the plant material undergoes extraction, which may follow different methods depending on the type of metabolite being sought or if a crude extract containing a mixture of metabolites is desired. These processes include maceration, Soxhlet extraction, or supercritical fluid extraction, among others.

After obtaining the extract, the solvent is removed through reduced-pressure separation using a rotary evaporator. Once most of the solvent is eliminated, the extract is collected and placed back into a drying oven until the solvent is completely removed (Verde-Star et al., 2016), as illustrated in Fig. 6.1.

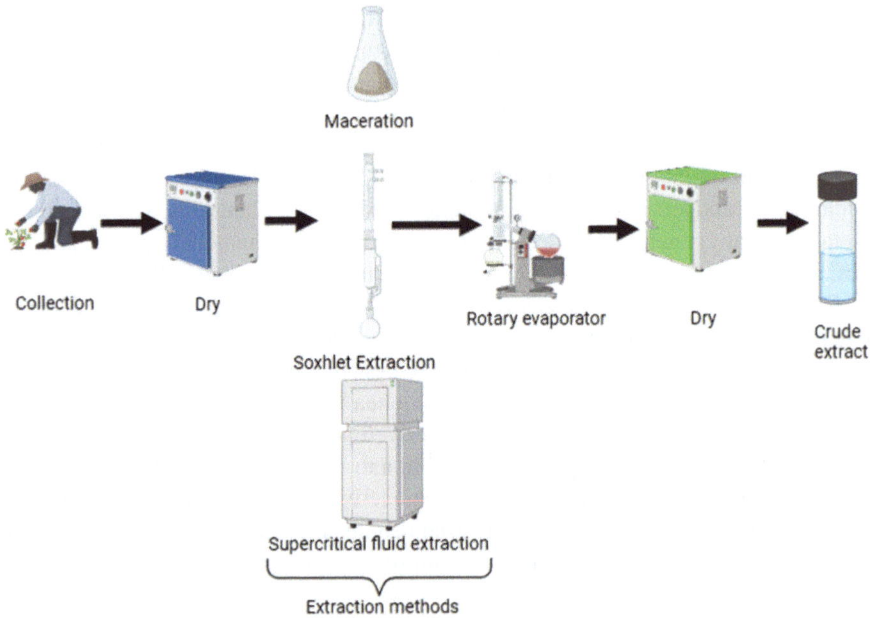

Fig. 6.1 Initial steps in obtaining plant extracts

Once the extract is obtained, there are various methodologies, which will be addressed throughout this book, to uncover its biological activity. Once potential is identified, different methods are employed to separate these extracts and elucidate the molecules responsible for the activity. Examples of plant-derived molecules that have significantly impacted disease treatment are listed in Table 6.1.

Owing to the relatively low cost of extraction systems, interest in the search for secondary metabolites or, alternatively, in modifying them to discover new health care applications has grown over time (Heya et al., 2023). Supercritical fluid extraction offers an advantage over maceration and Soxhlet extraction by using a gas as the eluent, thereby reducing the volume of chemical solvents required and contained for final disposal. However, a drawback of this system is its high cost (Agustina et al., 2024).

It is important to note that the pursuit of new applications for secondary metabolites involves the multidisciplinary collaboration of researchers, who must integrate diverse perspectives within their fields of study. This approach ensures reduced uncertainty in interpreting results. Consequently, collaboration among public and private universities, non-academic institutions, and national science and technology systems is critical for advancing scientific and technological development (Pereira et al., 2023).

In summary, plants are a source of natural products known as secondary metabolites, which exhibit a wide range of biological activities that impact human daily life. Their beneficial role extends beyond nutrition to include applications in human health, making them a valuable resource. However, the search for these metabolites

Table 6.1 Examples of the main natural products used in medicine

Name	Structure	Botanical source	Biological potential	References
Atropine	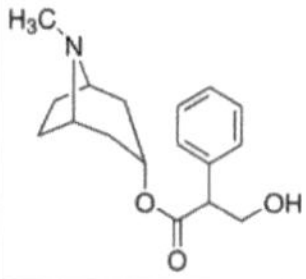	*Atropa belladonna* L.	Anticholinergic	Mehta et al. (2018)
Camptothecin	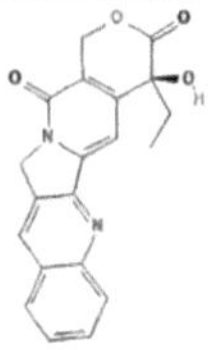	*Camptotheca acuminata* Decne.	Anticancer	Ghanbari-Movahed et al. (2021)
Digitoxin	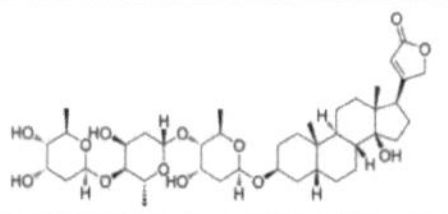	*Digitalis purpurea* L.	Cardiotonic	Anderson et al. (2012)
Morphine	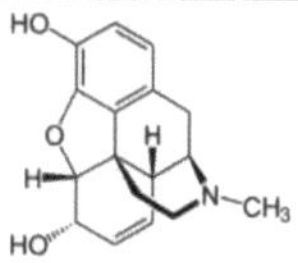	*Papaver somniferum* L.	Analgesic	Krames and Harb (2009)
Taxol	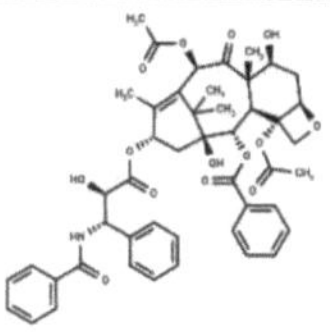	*Taxus brevifolia* Nutt.	Anticancer	Lim et al. (2022)
Thymol	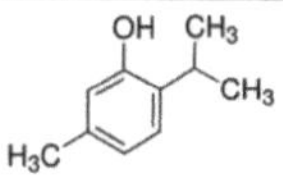	*Thymus vulgaris* L.	Antifungal	Marchese et al. (2016)

must employ specific methodologies to prevent the indiscriminate use of plant species (which risks driving them to extinction) and to minimize the generation of environmentally harmful byproducts. Such efforts should prioritize interdisciplinary collaboration to reduce uncertainty in result interpretation and foster sustainable scientific progress.

References

Agustina, G., Matías, S., Lucia, C., & Santiago Daniel, P. (2024). Fluidos supercríticos Innovación y sostenibilidad en la extracción de compuestos naturales. *Indufarma*, 34–37. http://hdl.handle.net/11336/252884.

Amit, M., Cohen, I., Marcovics, A., Muklada, H., Glasser, T. A., Ungar, E. D., & Landau, S. Y. (2013). Self-medication with tannin-rich browse in goats infected with gastro-intestinal nematodes. *Veterinary Parasitology, 198*(3–4), 305–311. https://doi.org/10.1016/j.vetpar.2013.09.019

Anderson, E. S., Lin, C.-H., Xiao, X., Stoilov, P., Burge, C. B., & Black, D. L. (2012). The cardiotonic steroid digitoxin regulates alternative splicing through depletion of the splicing factors SRSF3 and TRA2B. *RNA, 18*(5), 1041–1049. https://doi.org/10.1261/rna.032912.112

Arana, M. D., Correa, A. L., & Oggero, A. J. (2014). El reino plantae: ¿Qué es una planta y cómo se clasifican?: Un cambio paradigmático. *Revista de Educación En Biología, 17*(1), 9–24.

Atanasov, A. G., Zotchev, S. B., Dirsch, V. M., Orhan, I. E., Banach, M., Rollinger, J. M., Barreca, D., Weckwerth, W., Bauer, R., Bayer, E. A., Majeed, M., Bishayee, A., Bochkov, V., Bonn, G. K., Braidy, N., Bucar, F., Cifuentes, A., D'Onofrio, G., Bodkin, M., et al. (2021). Natural products in drug discovery: Advances and opportunities. *Nature Reviews Drug Discovery, 20*(3), 200–216. https://doi.org/10.1038/s41573-020-00114-z

Avery, L. M., & Hains, B. J. (2017). Oral traditions: A contextual framework for complex science concepts—Laying the foundation for a paradigm of promise in rural science education. *Cultural Studies of Science Education, 12*(1), 129–166. https://doi.org/10.1007/s11422-016-9761-5

Balunas, M. J., & Kinghorn, A. D. (2005). Drug discovery from medicinal plants. *Life Sciences, 78*(5), 431–441. https://doi.org/10.1016/j.lfs.2005.09.012

Benkhnigue, O., Chaachouay, N., Khamar, H., El Azzouzi, F., Douira, A., & Zidane, L. (2022). Ethnobotanical and ethnopharmacological study of medicinal plants used in the treatment of anemia in the region of Haouz-Rehamna (Morocco). *Journal of Pharmacy & Pharmacognosy Research, 10*(2), 279–302. https://doi.org/10.56499/jppres21.1196_10.2.279

Chaachouay, N., & Zidane, L. (2024). Plant-derived natural products: A source for drug discovery and development. *Drugs and Drug Candidates, 3*(1), 184–207. https://doi.org/10.3390/ddc3010011

d'Avigdor, E., Wohlmuth, H., Asfaw, Z., & Awas, T. (2014). The current status of knowledge of herbal medicine and medicinal plants in fiche. *Ethiopia. Journal of Ethnobiology and Ethnomedicine, 10*(1), 38. https://doi.org/10.1186/1746-4269-10-38

De la Fuente, M. F., Souto, A., Albuquerque, U. P., & Schiel, N. (2022). Self-medication in nonhuman primates: A systematic evaluation of the possible function of the use of medicinal plants. *American Journal of Primatology, 84*(11). https://doi.org/10.1002/ajp.23438

Diamond, A. M. (2006). The relative success of private funders and government funders in funding important science. *European Journal of Law and Economics, 21*(2), 149–161. https://doi.org/10.1007/s10657-006-6647-0

Fan, T.-P., Deal, G., Koo, H.-L., Rees, D., Sun, H., Chen, S., Dou, J.-H., Makarov, V. G., Pozharitskaya, O. N., Shikov, A. N., Kim, Y. S., Huang, Y.-T., Chang, Y. S., Jia, W., Dias, A., Wong, V. C., & Chan, K. (2012). Future development of global regulations of Chinese herbal products. *Journal of Ethnopharmacology, 140*(3), 568–586. https://doi.org/10.1016/j.jep.2012.02.029

García-Aguilar, M. C., Luévano-Esparza, J., & De la Cueva, H. (2017). La fauna nativa de México en riesgo y la NOM-059: ¿Están todos los que son y son todos los que están? *ACTA ZOOLÓGICA MEXICANA (N.S.), 33*(2), 188–198. https://doi.org/10.21829/azm.2017.3321060

Ghanbari-Movahed, M., Kaceli, T., Mondal, A., Farzaei, M. H., & Bishayee, A. (2021). Recent advances in improved anticancer efficacies of Camptothecin Nano-formulations: A systematic review. *Biomedicine, 9*(5). https://doi.org/10.3390/biomedicines9050480

Hardy, K. (2020). Paleomedicine and the evolutionary context of medicinal plant use. *Revista Brasileira de Farmacognosia, 31*(1), 1–15. https://doi.org/10.1007/s43450-020-00107-4

He, S. (2013). Chinese herbal dose in ancient and modern times: A comparative study. *Journal of Traditional Chinese Medicine, 33*(2), 268–271. https://doi.org/10.1016/S0254-6272(13)60138-7

Heya, M. S., García-Ponce, R., Soto, B. A. M., Verde-Star, M. J., Soto-Domínguez, A., García-Hernandez, D. G., Saucedo-Cárdenas, O., Hernández-Salazar, M., & Guillén-Meléndez, G. A. (2023). Green alternatives in treatment of liver diseases: The challenges of traditional medicine and green Nanomedicine. *Chemistry & Biodiversity, 20*(10). https://doi.org/10.1002/cbdv.202300463

Hu, X., Wang, M., Cai, F., Liu, L., Cheng, Z., Zhao, J., Zhang, Q., & Long, C. (2024). A comprehensive review of medicinal Toxicodendron (Anacardiaceae): Botany, traditional uses, phytochemistry and pharmacology. *Journal of Ethnopharmacology, 318*, 116829. https://doi.org/10.1016/j.jep.2023.116829

Jalota, K., Sharma, V., Agarwal, C., & Jindal, S. (2024). Eco-friendly approaches to phytochemical production: Elicitation and beyond. *Natural Products and Bioprospecting, 14*(1), 5. https://doi.org/10.1007/s13659-023-00419-7

Krames, E. S., & Harb, M. (2009). The rational use of intrathecal opioid analgesics. In *Neuromodulation* (pp. 441–455). Elsevier. https://doi.org/10.1016/B978-0-12-374248-3.00034-3

Kumar Shakya, A., & Arvind Kumar Shakya, C. (2016). Medicinal plants: Future source of new drugs. *International Journal of Herbal Medicine, 4*(4), 59–64. https://doi.org/10.13140/RG.2.1.1395.6085

Lim, P. T., Goh, B. H., & Lee, W.-L. (2022). Taxol: Mechanisms of action against cancer, an update with current research. In *Paclitaxel* (pp. 47–71). Elsevier. https://doi.org/10.1016/B978-0-323-90951-8.00007-2

Marchese, A., Orhan, I. E., Daglia, M., Barbieri, R., Di Lorenzo, A., Nabavi, S. F., Gortzi, O., Izadi, M., & Nabavi, S. M. (2016). Antibacterial and antifungal activities of thymol: A brief review of the literature. *Food Chemistry, 210*, 402–414. https://doi.org/10.1016/j.foodchem.2016.04.111

Mehta, P., Bothiraja, C., Mahadik, K., Kadam, S., & Pawar, A. (2018). Phytoconstituent based dry powder inhalers as biomedicine for the management of pulmonary diseases. *Biomedicine & Pharmacotherapy, 108*, 828–837. https://doi.org/10.1016/j.biopha.2018.09.094

Muñoz-Romo, M., Cohen, G., & Page, R. A. (2025). Actual folivory in bats: Whole leaves as a formerly dismissed trophic item and a potential avenue for self-medication. *Behaviour, 162*(2), 173–190. https://doi.org/10.1163/1568539X-bja10297

Pereira, R. M., Borini, F. M., & Oliveira, M. de M. (2023). Technological upgrade in emerging markets: The importance of Interorganizational cooperation for product and process upgrades. International Journal of Innovation and Technology Management, 20(08). https://doi.org/10.1142/S0219877023500517.

Rodrigues, M. J. (2025). Natural products: Sources and applications. *Applied Sciences, 15*(4), 1716. https://doi.org/10.3390/app15041716

Thelwall, M., Simrick, S., Viney, I., & Van den Besselaar, P. (2023). What is research funding, how does it influence research, and how is it recorded? *Key dimensions of variation. Scientometrics, 128*(11), 6085–6106. https://doi.org/10.1007/s11192-023-04836-w

Verde-Star, M. J., García-González, S., & Rivas-Morales, C. (2016). Metodología científica para el estudio de plantas medicinales. In *Investigación en plantas de importancia médica* (pp. 1–40). https://doi.org/10.3926/oms.335

Chapter 7
Small Molecules from Microalgae as an Emerging Sustainable Alternative for Disease Management

Daniela Treviño-Almaguer, Bryan Alejandro Espinosa-Rodríguez, Isaías Balderas-Rentería, and Jose Luis González-Llerena

1 Introduction

Microalgae are unicellular photosynthetic microorganisms that inhabit aquatic ecosystems, in both freshwater and marine environments (Caroppo & Pagliara, 2022; Dolganyuk et al., 2020; Thoré et al., 2023). Despite their microscopic size, their biological and ecological significance is immense as they form the base of aquatic food chains and are estimated to produce around 50% of the Earth's oxygen (Hu et al., 2023). By fixing atmospheric carbon dioxide through photosynthesis, they contribute significantly to the global carbon cycle and climate balance. Their widespread abundance and foundational ecological function position them not only as pivotal environmental agents but also as a promising and increasingly recognized source of bioactive compounds with significant therapeutic potential (Dolganyuk et al., 2020; Ighalo et al., 2022; Sarıtaş et al., 2024; Tsvetanova & Yankov, 2022).

In recent decades, interest in microalgae has grown as a sustainable alternative for disease management. Unlike many traditional drug sources, microalgae can be rapidly cultivated using sunlight as an energy source, with relatively low requirements and a minimal carbon footprint (Onyeaka et al., 2021; Sarker & Kaparaju, 2024; Srimongkol et al., 2022; Yu et al., 2024). These "cellular biorefineries" convert CO_2 into biomass rich in medically valuable metabolites, all without competing for arable land or generating the toxic waste associated with conventional chemical synthesis. Their rapid growth, scalability, and the absence of human pathogens in

D. Treviño-Almaguer · I. Balderas-Rentería · J. L. González-Llerena (✉)
Laboratory of Molecular Pharmacology and Biological Models, School of Chemistry, Universidad Autónoma de Nuevo León, Nuevo León, Mexico
e-mail: jgonzalezl@uanl.edu.mx

B. A. Espinosa-Rodríguez
Laboratory of Biopharmacy, School of Chemistry, Universidad Autónoma de Nuevo León, Nuevo León, Mexico

M. S. Heya (ed.), *Natural Alternatives and Nanotechnology Applied to the One Health Approach*, https://doi.org/10.1007/978-3-032-03675-9_7

their cultivation make them particularly attractive for the sustainable production of pharmaceuticals (Okeke et al., 2022; Olguín et al., 2022; Uma et al., 2023).

The search for more sustainable therapies is also driven by the limitations of traditional pharmacotherapy. Issues such as emerging antibiotic resistance, high costs of drug synthesis and distribution, and the side effects of chronic treatments highlight the need for novel approaches. For instance, prolonged use of nonsteroidal anti-inflammatory drugs (NSAIDs) can lead to severe gastrointestinal, renal, and hepatic adverse effects (Ghosh et al., 2022; Mhadhebi et al., 2014; Sagnia et al., 2014), which has fueled the search for safer natural alternatives. In this context, microalgae emerge as promising candidates. A growing body of evidence suggests that bioactive compounds derived from algal extracts may offer comparable pharmacological efficacy to NSAIDs, antibiotics, and other conventional drugs, while exhibiting a more favorable safety profile and reduced systemic toxicity (Fernando et al., 2016; Obluchinskaya et al., 2022). As inflammation, oxidative stress, and infections are involved in numerous human pathologies, microalgae-derived molecules offer an integrated and eco-friendly approach to disease management, aligned with the urgent need to transition toward more sustainable medicine (Choo et al., 2020; Melloni et al., 2023; Pereira et al., 2024; Vignaud et al., 2023).

Accordingly, microalgae embody a rare convergence of pharmacological potential and environmental sustainability, offering advantages that few biological sources can parallel. The sections that follow explore, in depth, their taxonomic and metabolic diversity, the principal classes of small bioactive molecules they synthesize, and the molecular pathways through which these compounds exert therapeutic effects across a range of diseases. This chapter also examines the ecological benefits of microalgae relative to conventional sources, alongside recent advances in biotechnological production, regulatory and market barriers, and emerging strategies aimed at positioning microalgae as a cornerstone of sustainable medicine in the twenty-first century.

2 Biodiversity and Bioactive Potential of Microalgae

Microalgae constitute a highly diverse polyphyletic group of photosynthetic organisms, encompassing both prokaryotic forms, such as cyanobacteria or "blue-green algae," and unicellular eukaryotes from various evolutionary lineages. Taxonomically, they span multiple phyla and genera with therapeutic potential. For example, within the green algae (Chlorophyta), *Chlorella* and *Dunaliella* are notable for their richness in carotenoid pigments (Pereira et al., 2024); *Haematococcus pluvialis* (another chlorophyte) is renowned for its high astaxanthin content (Ayub et al., 2025); diatoms such as *Phaeodactylum* produce fucoxanthin and omega-3 fatty acids

(Su et al., 2023; Wu et al., 2010); and among cyanobacteria, *Spirulina* (genus *Arthrospira*) stands out owing to its high protein and phycocyanin content (Ayub et al., 2025).

The breadth of this taxonomic diversity is mirrored in the remarkable metabolic plasticity of microalgae, with each lineage evolving distinct biosynthetic pathways tailored to specific ecological niches ranging from hypersaline lakes to extreme marine habitats. These evolutionary adaptations enable the synthesis of structurally diverse and pharmacologically active compounds, constituting a vast reservoir of bioactive metabolites with clinical potential (Dolganyuk et al., 2020; Osathanunkul et al., 2025; Rojas-Villalta et al., 2024).

One of the defining features of microalgal bioactivity is their capacity to accumulate high levels of secondary metabolites under environmental stress. Exposure to factors such as elevated salinity, nutrient limitation, or high solar radiation triggers the production of protective compounds (Kolackova et al., 2023). For instance, *Dunaliella salina* can accumulate more than 10% of its biomass as β-carotene under saline or light stress, turning the salt ponds where it grows a vivid orange (Aditi et al., 2025). Similarly, *Haematococcus* forms red cysts rich in astaxanthin, serving as a powerful antioxidant defense (Yao et al., 2020; Zhang et al., 2018). Certain cyanobacteria produce unusual amino acids and cyclic peptides as protective agents against predators or UV radiation (e.g., micropeptins, UV-screening metabolites such as scytonemin) (Leão et al., 2010; Wada et al., 2013). This ability to synthesize rare or highly concentrated molecules often surpasses that of terrestrial plants or fungi, positioning microalgae as a valuable source of small bioactive compounds that are difficult to obtain through other means (Parmar et al., 2023).

Another noteworthy feature is the nutritional richness and safety profile of several microalgae, which has facilitated their adoption as dietary supplements and serves as an indicator of their biocompatibility. *Spirulina,* a filamentous cyanobacterium, has been used ancestrally as food, for example, by Aztec cultures, and more recently was selected by NASA as a dietary supplement for astronauts owing to its high protein (up to 70% of dry weight) and vitamin content (Karkos et al., 2011). This organism provides B-complex vitamins (including B_{12}), provitamin A (β-carotene), and essential fatty acids, with minimal adverse effects reported. In fact, *Spirulina* is Generally Recognized As Safe (GRAS) by the FDA, following numerous toxicological studies demonstrating its safety (Marles et al., 2011).

Likewise, species such as *Chlorella vulgaris* are marketed globally as superfoods owing to their comprehensive profile of essential micronutrients, including vitamins A, C, E, B-complex, D_2, iron, and others, often in higher concentrations than conventional vegetables (Bito et al., 2020; Sandgruber et al., 2021). The presence of vitamin B_{12} and vitamin D_2 in *Chlorella* nutrients, typically absent in higher plants, highlights the biochemical uniqueness of microalgae and their potential advantage as a comprehensive nutraceutical source.

3 Bioactive Molecules Derived from Microalgae

Microalgae synthesize a wide array of low-molecular-weight organic compounds with biological activity, many of which belong to classes of secondary metabolites also found in plants, yet with distinctive structural variations. The bioactive small molecules produced by microalgae can be broadly grouped into six main chemical categories, each with distinct structural characteristics and therapeutic mechanisms. These include carotenoids, which are isoprenoid pigments; polyunsaturated fatty acids (PUFAs), known for their cardiometabolic benefits; bioactive peptides, and small proteins with functions such as enzyme inhibition and neuroprotection; phenolic compounds that exhibit antioxidant and anti-inflammatory properties; polysaccharides, particularly those with sulfated structures that show immunomodulatory and antiviral effects; and finally, alkaloids along with other nitrogen-containing metabolites, many of which display cytotoxic or neuroactive potential. The following sections explore each of these classes, highlighting key examples, and their relevance to disease intervention.

3.1 Carotenoids

Among the most visually striking and biologically versatile metabolites produced by microalgae are carotenoids, which are lipophilic terpenoid pigments responsible for the yellow, orange, and red hues observed in many photosynthetic organisms. Beyond their primary roles in light harvesting and photoprotection within algal cells, these compounds have attracted considerable attention for their potent antioxidant activity and broad-spectrum health benefits. In the context of human nutrition and therapeutic development, carotenoids derived from microalgae are especially notable owing to their capacity to function as both essential micronutrients and pharmacologically active agents. Within this class, the most studied and commercially relevant examples are astaxanthin, β-carotene, and fucoxanthin, each characterized by distinctive molecular structures and specific biological functions that contribute to their nutraceutical and medicinal value (Ambati et al., 2014; Baeza-Morales et al., 2024; Bjørklund et al., 2022; Gholipour-Varnami et al., 2025; Karpiński et al., 2021; Oliyaei et al., 2023; Shanaida et al., 2025).

Astaxanthin is a reddish xanthophyll produced in large quantities by the green microalga *Haematococcus pluvialis*. It has a structure similar to that of β-carotene but with oxygenated functional groups (ketone and hydroxyl) on its terminal rings, granting it exceptional antioxidant capacity by enabling it to scavenge free radicals in both polar and nonpolar regions of cell membranes (Si & Zhu, 2022) Thanks to these antioxidant, anti-inflammatory, and anti-apoptotic properties, astaxanthin has demonstrated wide-ranging benefits in disease models: it protects against oxidative damage in ocular and skin tissues, improves cardiovascular parameters, and favorably modulates immune responses (Dang et al., 2024; Si & Zhu, 2022; Wolf et al.,

2021). A particularly noteworthy feature is its ability to cross the blood–brain barrier, prompting interest in its use for neurological disorders. Astaxanthin has demonstrated neuroprotective properties in experimental models of neurological disorders, including Alzheimer's disease, Parkinson's disease, cerebral ischemia, and other pathologies affecting the nervous system (Si & Zhu, 2022).

Another major carotenoid with well-established health benefits is β-carotene. This orange-colored provitamin A compound is abundantly accumulated by *Dunaliella salina* reaching up to 14% of dry biomass and other halotolerant green algae (Wolf et al., 2021). It is used as a dietary supplement to support visual, skin, and immune health (Aditi et al., 2025). Its primary actions include antioxidant activity and serving as a precursor of retinal/retinoids, which are essential for vision and cellular differentiation (Baeza-Morales et al., 2024; Cheon et al., 2023; Shanaida et al., 2025; Sher et al., 2018).

Distinct from carotenoids found in green algae, fucoxanthin is specific to brown algae and diatoms, such as *Phaeodactylum tricornutum*. This brown-green pigment contributes to photosynthetic efficiency under blue-green light (Lee et al., 2021). It exhibits well-studied anti-obesity and antitumor properties in murine models, reduces adipogenesis, enhances fat oxidation by inducing the expression of uncoupling protein 1 (UCP1) in white adipose tissue, enhancing lipid oxidation, and reducing fat accumulation (Maeda, 2015; Mumu et al., 2022). It also exerts anti-inflammatory effects by inhibiting the release of pro-inflammatory cytokines and blocking activation of the NLRP3 inflammasome in immune cells (Gallego et al., 2022; Lee et al., 2021).

Collectively, microalgal carotenoids function as broad-spectrum antioxidants that protect cellular components, including lipids, nucleic acids, and proteins from oxidative damage (Gallego et al., 2022). They also modulate key signaling pathways, such as NF-κB and transcriptional regulatory networks (Koo et al., 2019), which underlie their anti-inflammatory, immunomodulatory, and chemopreventive effects against cancer (Dang et al., 2024; Si & Zhu, 2022).

3.2 *Polyunsaturated Fatty Acids*

Many marine microalgae are rich in long-chain PUFAs from the omega-3 and omega-6 families. Genera such as *Schizochytrium*, *Nannochloropsis*, and *Isochrysis* produce eicosapentaenoic acid (EPA, C20:5 ω-3) and docosahexaenoic acid (DHA, C22:6 ω-3), the same beneficial omega-3 fatty acids found in fish oil. These fatty acids have well-documented cardioprotective, anti-inflammatory, and neurodevelopmental effects in humans (Ganuza et al., 2024; Karrar et al., 2024; Ryckebosch et al., 2014). Their intake helps to reduce plasma triglyceride levels, improve cholesterol profiles, enhance endothelial function, and lower the risk of arrhythmias, collectively reducing cardiovascular events (Adarme-Vega et al., 2012).

At the molecular level, EPA and DHA serve as precursors to anti-inflammatory eicosanoids and resolvins, counteracting the production of pro-inflammatory prostaglandins and leukotrienes derived from arachidonic acid (Ganuza et al., 2024; Karrar et al., 2024; Ryckebosch et al., 2014). Moreover, their incorporation into neuronal membranes supports synaptic plasticity and protects against neurodegenerative processes (Crupi et al., 2013).

Microalgae represent a sustainable source of these omega-3 fatty acids, in contrast to extractive fishing practices. Algal oils are already used as alternatives to fish oil in human nutrition (e.g., dietary supplements, infant formulas) and aquaculture feeds, providing EPA and DHA free of contaminants such as dioxins and heavy metals (Santigosa et al., 2021). In addition to omega-3 s, certain cyanobacteria produce gamma-linolenic acid (C18:3 ω-6), an anti-inflammatory fatty acid found in *Spirulina*, and other microalgae synthesize arachidonic acid (C20:4 ω-6), which plays important roles in brain function (Sajilata et al., 2008; Su et al., 2016).

3.3 Bioactive Peptides and Proteins

Microalgae, being organisms with a high protein content, are capable of producing pharmacologically relevant peptides, either through normal metabolism or via enzymatic hydrolysis of their proteins. Notable examples include angiotensin-converting enzyme (ACE) inhibitory peptides derived from *Chlorella* and *Spirulina*. In studies, enzymatic hydrolysis of *Chlorella* proteins released tripeptides with potent antihypertensive activity, capable of inhibiting ACE at nanomolar levels (Li et al., 2021; Münzel & Daiber, 2019; Suo et al., 2022, 2024; Xie, Chen, et al., 2018a). Several of these algal peptides have been shown to reduce blood pressure in hypertensive rats, acting similarly to conventional ACE inhibitors (Zheng et al., 2017). Their short sequences and low molecular weight enhance absorption and bioactivity.

Spirulina also contains peptides with antioxidant and metal-chelating activity, and a specific 9-amino-acid peptide (referred to as SP6) has demonstrated neuroprotective effects in vitro (Bortolini et al., 2022; Carrizzo et al., 2019; Garbuzova-Davis et al., 2010). Another notable protein is phycocyanin, a blue pigment-protein found in cyanobacteria, which constitutes around 15% of *Spirulina*'s biomass. Phycocyanin has shown antioxidant properties (scavenging peroxyl radicals), anti-inflammatory effects (inhibiting histamine release by mast cells), and even anticancer activity by inducing apoptosis in tumor cells (Fernandes et al., 2023; Jiang et al., 2017; Remirez et al., 2002). Part of *Spirulina*'s anti-arthritic effect is attributed to the ability of phycocyanin to reduce lytic enzymes and pro-inflammatory cytokines in the joint (Remirez et al., 2002).

Thus, algal proteins are a valuable source of peptides with antihypertensive, antioxidant, immunomodulatory, and anticancer properties, expanding the therapeutic repertoire beyond typical secondary metabolites (Fernandes et al., 2023; Jiang et al., 2017). Targeted production of these peptides through controlled hydrolysis or recombinant expression in microalgae represents an emerging field in both food and pharmaceutical biotechnology.

3.4 Phenolic Compounds

Although in lower quantities than higher plants, microalgae also synthesize phenolic compound molecules containing hydroxylated aromatic rings with notable biological activity. Certain chlorophytes and diatoms produce flavonoids and phenolic acids (such as gallic, p-coumaric, and caffeic acids), which contribute to their antioxidant capacity (Ferdous et al., 2025). Some cyanobacteria generate unique phenolic metabolites, including mycosporine-like amino acids (MAAs), which are modified amino acids with potent photoprotective properties (Punchakara et al., 2023). These compounds, found in species such as *Porphyridium cruentum* and other microalgae, absorb UV radiation, acting as "molecular sunscreens" that prevent oxidative damage (Cao et al., 2016).

Studies suggest that MAAs may also inhibit skin-degrading enzymes and reduce inflammation, pointing to potential applications in cosmetics and dermatology as natural photoprotective agents (Álvarez-Gómez et al., 2019). Additionally, blue–green algae synthesize halogenated phenolic compounds with antimicrobial activity (Bhuyar et al., 2020).

Overall, phenolics from microalgae function as antioxidants and chemoprotective agents, neutralizing free radicals and preventing lipid peroxidation. Similar to plant-derived polyphenols (e.g., resveratrol, quercetin), their therapeutic mechanisms may include activation of cytoprotective pathways (e.g., Nrf2) and modulation of pro-inflammatory cellular signaling (Proença et al., 2025; Wu et al., 2022).

It is important to note that research on microalgal phenolics is relatively recent, and many of these compounds likely remain uncharacterized. Nevertheless, their presence reinforces the concept that microalgae offer a chemically rich profile comprising pigments, lipids, peptides, and phenolics that combines synergistic health-promoting activities (Riccio & Lauritano, 2019).

3.5 Polysaccharides

Numerous microalgae produce polysaccharides, both structural (cell-wall components) and extracellular, several of which exhibit notable bioactive properties. A particularly important class are the sulfated polysaccharides secreted by marine microalgae (Wei et al., 2022). For instance, the red microalga *Porphyridium* releases a sulfated phycocolloid with antiviral and immunostimulatory effects. These sulfated polysaccharides structurally mimic glycosaminoglycans (such as heparan sulfate) found on cell surfaces, enabling them to block viral adsorption to host cells. Algal polysaccharides have shown antiviral activity against enveloped viruses such as herpes simplex, dengue, and more recently, coronaviruses, by interfering with virus-cell binding and enhancing the innate immune response (Ozório et al., 2024; Panggabean et al., 2022).

In addition, these complex sugars modulate immune activity by stimulating macrophages, increasing cytokine release, and activating natural-killer (NK) cells, thereby promoting a general antiviral state in the host (Wei et al., 2022). Another noteworthy polysaccharide is paramylon, a β-1,3-glucan stored in the photosynthetic protist *Euglena*, with a structure similar to yeast-derived beta-glucans (Nagayama et al., 2020). Studies suggest that paramylon might act as an immunomodulator by activating pattern recognition receptors (such as dectin-1), triggering defensive immune responses with potential applications in immunotherapy and vaccine adjuvants (Kanjan et al., 2017; Ryan et al., 2023; Yasuda et al., 2020).

Additionally, polysaccharides from *Chlorella* and *Spirulina* have been shown to stimulate antibody production and lymphocyte activity, enhancing immune responses in animal models (Riccio & Lauritano, 2019). Other microalgae produce nonsulfated polysaccharides with prebiotic effects supporting beneficial gut microbiota—and local anti-inflammatory actions in the gastrointestinal tract (Gotteland et al., 2020; Patel et al., 2021).

3.6 Alkaloids and Other Nitrogen-Containing Molecules

Some microalgae and cyanobacteria produce alkaloid compounds, heterocyclic nitrogen-based metabolites that exhibit potent biological activity. For example, indole and brominated alkaloids isolated from marine cyanobacteria have demonstrated selective cytotoxicity against cancer-cell lines. A prominent case is apratoxin A, a peptide–polyketide hybrid produced by the cyanobacterium *Lyngbya*, which exhibits strong anticancer activity by inhibiting growth factor signaling pathways; it is currently under investigation as a lead compound for the development of antitumor drugs(Gutiérrez et al., 2008; Liu & Rein, 2010; Luesch et al., 2001).

Other examples include anatoxin-a, a neurotoxic cholinergic agonist alkaloid produced by *Dolichospermum*. Although toxic, anatoxin-a has been used as a pharmacological tool in neurological studies (Chernova et al., 2017). In green microalgae, the production of true alkaloids is less common, but these organisms do synthesize amino sugars, polyamines, and pyrrolidine derivatives that may be considered part of this broad class (Jantaro & Kanwal, 2017; Yoon et al., 2025). Owing to their potent bioactivity, many of these compounds are being studied as lead compounds for the synthesis of therapeutic analogs.

In addition, the nitrogen metabolism of microalgae yields other molecules of pharmacological interest. *Spirulina*, for instance, contains phycocyanobilin, a linear tetrapyrrole structurally similar to bilirubin, which exerts antioxidant effects via the heme oxygenase-1 (HO-1) pathway (Strasky et al., 2013). Microalgae have also been found to produce polyamines such as spermidine and spermine, which may have cellular anti-aging effects (Minois, 2014).

Although microalgal alkaloids are neither as abundant nor as well-characterized as those from medicinal plants, they represent a promising area for the discovery of new pharmacological agents, some of which exhibit novel mechanisms of action (e.g., proteasome inhibitors, ion-channel modulators) (Efimova et al., 2020; Tsukamoto et al., 2010). Careful research is needed to isolate these compounds in sufficient quantities and evaluate their toxicity, as many are highly potent. Nonetheless, their presence broadens the spectrum of bioactive microalgal chemistry, reinforcing the idea that nearly every class of pharmacologically useful secondary metabolite, whether carotenoid, lipid, phenolic, polysaccharide, or alkaloid, may have a microalgal analog.

4 Molecular Mechanisms Involved in Disease Management

The therapeutic effects of microalgae-derived molecules are supported by a variety of molecular mechanisms that positively interfere with pathogenic pathways in disease processes. Among the most important are their antioxidant and anti-inflammatory actions, the regulation of cellular pathways associated with oxidative stress and chronic inflammation, modulation of signaling pathways involved in cell proliferation (notably in cancer) and metabolism, as well as emerging influences on epigenetic regulation. The key mechanisms are outlined below.

4.1 Antioxidant Activity of Microalgal Compounds

Microalgal bioactives exhibit potent antioxidant properties through both direct and indirect mechanisms. Carotenoids such as astaxanthin, β-carotene, and fucoxanthin neutralize reactive oxygen and nitrogen species (ROS/RNS) via their conjugated double bonds, effectively stabilizing free radicals and halting lipid peroxidation in cell membranes. Notably, astaxanthin has shown superior antioxidant capacity compared with classical antioxidants such as vitamin C and α-tocopherol (Aditi et al., 2025; Si & Zhu, 2022).

Additionally, phycocyanin from *Spirulina* scavenges peroxyl radicals and singlet oxygen, reducing systemic oxidative stress. Beyond direct scavenging, several microalgal metabolites activate the Nrf2 signaling pathway, a master regulator of antioxidant defense. Nrf2 upregulates cytoprotective enzymes such as heme oxygenase-1, superoxide dismutase, catalase, and glutathione S-transferases, enhancing cellular resilience (Strasky et al., 2013; Wu et al., 2022).

Compounds such as astaxanthin, phycocyanin, and algal polyphenols have been documented to trigger Nrf2 activation, contributing to reduced oxidative damage in inflammatory or ischemic conditions (Fernández-Rojas et al., 2025; Proença et al., 2025; Wu et al., 2022). This dual mechanism, free radical scavenging and endogenous enzyme induction, positions microalgal antioxidants as promising tools in the prevention and management of oxidative-stress-linked diseases such as atherosclerosis, Alzheimer's, and Parkinson's (Matin et al., 2024).

4.2 Anti-inflammatory Activity

Chronic inflammation underlies numerous diseases, including arthritis, inflammatory bowel disease, metabolic syndrome, and cancer. This process is sustained by excessive production of pro-inflammatory mediators such as cytokines (e.g., TNF-α, IL-1β, IL-6), eicosanoids (e.g., prostaglandins), and nitric oxide. Microalgal compounds, particularly polyphenols, carotenoids (such as fucoxanthin), and polysaccharides, have shown significant potential in modulating these responses.

A primary mechanism involves the inhibition of pro-inflammatory signaling pathways in immune cells. Microalgal extracts can suppress the activation of nuclear factor NF-κB, a transcription factor controlling the expression of inflammatory genes such as COX-2 and iNOS. This results in reduced levels of cytokines and prostaglandins (Matin et al., 2024). Additionally, microalgal compounds influence MAPK pathways (e.g., p38, ERK, JNK), reducing phosphorylation events that trigger inflammatory mediator synthesis. Fucoxanthin, for instance, inhibits p38 MAPK activation in stimulated macrophages, thereby lowering IL-1β production. Inhibition of the PI3K/Akt pathway by algal molecules also contributes to inflammation resolution. Moreover, microalgae provide omega-3 fatty acids such as EPA, which shift eicosanoid synthesis toward less inflammatory series (e.g., series-3 prostaglandins, series-5 leukotrienes) and promote resolvins that actively resolve inflammation (Cheon et al., 2023; Matin et al., 2024; Nabil-Adam et al., 2023).

In vivo studies confirm these effects. *Spirulina* and *Chlorella* extracts reduce leukocyte infiltration and inflammatory markers in damaged tissues. In murine arthritis models, *Spirulina* decreased lysosomal β-glucuronidase and preserved cartilage integrity (Lee et al., 2021). In colitis models, *Spirulina* attenuated cytokine release and oxidative stress, improving intestinal histology and even modulating gene methylation linked to inflammation (de Brito et al., 2020; Matin et al., 2024; Naselli et al., 2025; Remirez et al., 2002).

4.3 Regulation of Cancer Pathways

The antiproliferative and pro-apoptotic properties of many microalgal compounds are closely linked to their ability to interfere with cell-cycle signaling cascades and tumor-cell survival mechanisms. Carotenoids such as astaxanthin and fucoxanthin induce apoptosis in cancer cells through both intrinsic (mitochondrial) and extrinsic pathways. Fucoxanthin has been shown to upregulate pro-apoptotic proteins (e.g., Bax) and caspases, while downregulating anti-apoptotic proteins (e.g., Bcl-2) in colon- and breast-cancer cell lines, triggering programmed cell death. Similarly, astaxanthin can induce cell-cycle arrest at the G_0/G_1 phase in malignant cells by modulating cyclins and cyclin-dependent kinases.

A key target in cancer is the PI3K/Akt/mTOR pathway, which is often overactivated to promote growth and resistance to apoptosis. Several microalgal extracts have demonstrated inhibition of Akt and mTOR, thereby sensitizing tumor cells to death. For example, sulfated polysaccharides from red microalgae have been shown to reduce Akt phosphorylation and increase AMPK activation in carcinoma-cell cultures, thereby altering tumor metabolism and suppressing proliferation (Riccio & Lauritano, 2019).

Additionally, microalgal metabolites can affect the tumor microenvironment. Certain cyanobacteria produce compounds such as curacin A and lagunamide, which inhibit tumor angiogenesis by blocking the vascular endothelial growth factor (VEGF) pathway. On the immunological front, microalgal polysaccharides and even whole *Spirulina* act as immunoadjuvants by enhancing the activity of NK cells and cytotoxic T lymphocytes, thereby indirectly aiding the immune system by eliminating cancer cells (Akao et al., 2009; Ferdous & Yusof, 2021; Masoumi et al., 2025; Vujović et al., 2025).

Both in vitro and in vivo studies support the ability of various microalgal compounds to inhibit tumor proliferation, induce apoptosis, and reduce invasiveness (Xie et al., 2018b). In leukemia and lymphoma models, algal extracts have successfully triggered malignant cell death and arrested cell-cycle progression (Ferdous & Yusof, 2021; Sharma et al., 2023). Although most antitumor evidence is still preclinical, these findings lay a strong foundation for developing algae-based adjuvant cancer therapies. Notably, many of these compounds exhibit selectivity by preferentially targeting cancer cells over normal cells, likely because of the activation of stress-related pathways that are more prominent in tumors (Vujović et al., 2025).

4.4 Modulation of Metabolic and Cardiovascular Pathways

In metabolic diseases such as type 2 diabetes, dyslipidemia, and obesity, bioactive compounds from microalgae exert effects across multiple molecular targets. Omega-3 fatty acids (EPA and DHA) improve insulin sensitivity in peripheral tissues by integrating into cell membranes and reducing low-grade chronic inflammation associated with insulin resistance. These fatty acids activate PPAR-γ in adipose tissue, promoting healthy adipogenesis and the secretion of adiponectin, a hormone that enhances glucose uptake and fatty-acid oxidation in muscle and liver, thereby improving glycemic control and lipid profiles.

Clinical trials have shown that supplementation with omega-3-rich microalgae reduces plasma triglycerides and VLDL cholesterol levels, similar to fish-oil supplementation (Adarme-Vega et al., 2012; Santigosa et al., 2021). Additionally, phycobiliproteins and polysaccharides from *Spirulina* and *Chlorella* appear to modulate key liver enzymes. *Spirulina* has been shown to inhibit HMG-CoA reductase, the rate-limiting enzyme in cholesterol synthesis, potentially explaining observed LDL cholesterol reductions in regular consumers.

A meta-analysis of 797 subjects reported that *Chlorella* supplementation significantly improved total cholesterol, LDL, systolic blood pressure, and fasting glucose levels, suggesting a broad, beneficial impact on metabolic syndrome (Bito et al.,

2020). Molecular studies indicate that *Chlorella* may upregulate hepatic LDL receptors, facilitating cholesterol clearance. Regarding glycemia, certain microalgal peptides can inhibit digestive enzymes (α-amylase, α-glucosidase), reducing carbohydrate absorption, whereas magnesium and B vitamins from *Spirulina* and *Chlorella* act as cofactors for insulin secretion and action. Moreover, the antioxidant effects of algal compounds protect vascular endothelium from hyperglycemia-induced oxidative stress, enhancing endothelial function and lowering cardiovascular risk in diabetes. In obesity, fucoxanthin promotes fat burning through UCP1 activation, inducing "browning" of white adipose tissue and increasing energy expenditure (Maeda, 2015; Mumu et al., 2022). This, combined with reduced adipose-derived cytokines, has led to less weight and fat gain in animal models.

Cardiovascular benefits are also observed through lipid-lowering, antihypertensive (via ACE-inhibitory peptides), and antithrombotic (via anti-inflammatory platelet modulation by omega-3 s) effects. For example, in adults with mild hypertension, *Spirulina* consumption reduced blood pressure, likely due in part to bioactive peptides with ACE inhibitory activity (Heo et al., 2017; Martínez-Sámano et al., 2018).

4.5 Epigenetic Pathway Modulation

A promising frontier in the study of natural compounds is their impact on epigenetics, defined as heritable modifications in gene expression that occur without changes to the underlying DNA sequence (Farheen et al., 2024; Gibney & Nolan, 2010). In this context, recent studies suggest that certain microalgae may exert epigenetic effects by modulating DNA methylation and possibly histone modifications (Bacova et al., 2020).

A pioneering finding involves the cyanobacterium *Aphanizomenon flos-aquae* (commonly known as Klamath blue-green algae), whose extract was shown to alter DNA methylation patterns in models of intestinal inflammation. In a study involving rats with colitis, administration of *Aphanizomenon* extract partially reversed the global DNA hypomethylation induced by chronic inflammation, suggesting that its compounds might promote the re-methylation and silencing of inflammatory genes (Naselli et al., 2025; Neupert et al., 2020). In cytokine-exposed Caco-2 intestinal-cell cultures, the same extract also modulated the methylation of promoter regions of key genes, thereby attenuating inflammatory activation. These findings were among the first to suggest that microalgae can influence DNA methylation in human or animal models, opening a novel dimension of inter-kingdom interaction. Such epigenetic capabilities imply that microalgal compounds may act by "reprogramming" the expression of disease-associated genes. Although the exact mechanisms remain unclear, it is hypothesized that polyphenols or metabolites such as phycocyanin might affect the activity of DNA methyltransferases or ten-eleven translocation enzymes (TETs), key regulators of methyl-group addition or removal in the genome (Naselli et al., 2025).

Furthermore, some carotenoids from microalgae have been shown to alter histone acetylation in tumor cells, promoting the expression of tumor suppressor genes. This form of epigenetic modulation is therapeutically attractive because it could induce more sustained and broad-spectrum effects by restoring healthy gene expression patterns in diseased cells (Ferrari et al., 2023; Naselli et al., 2025; Steadman et al., 2020). Although still a nascent field, current evidence positions microalgae as potential epigenetic nutraceutical agents—capable of influencing the genome's regulatory machinery and inducing cellular states more resistant to chronic stress and inflammation. Further investigation into which specific metabolites are responsible for these effects (e.g., certain flavonoids, alkaloids, or even vitamin B_{12}, which plays a role in methylation processes) will be essential to harness this therapeutic potential.

5 Specific Therapeutic Applications of Microalgae Compounds

Bioactive compounds derived from microalgae have shown considerable promise in the treatment and prevention of a broad spectrum of diseases, including inflammatory conditions, cancer, metabolic, and neurodegenerative disorders, as well as infectious diseases. An overview of recent evidence illustrates the therapeutic potential of these molecules across multiple clinical domains.

In the context of chronic inflammatory diseases, preclinical models of arthritis have demonstrated that supplementation with *Spirulina platensis* significantly reduces joint inflammation (Ali et al., 2015; Yang et al., 2025). This effect has been attributed to the antioxidant and anti-inflammatory actions of its bioactive components, which contribute to preserving cartilage integrity and limiting leukocyte infiltration (Yang et al., 2025). Furthermore, in a randomized, double-blind, placebo-controlled trial involving patients with mild to moderate ulcerative colitis, daily administration of *Spirulina* (1 g for 8 weeks) improved systemic antioxidant capacity and enhanced health-related quality of life, suggesting beneficial effects in human inflammatory bowel disease (Moradi et al., 2024).

Regarding cancer, several microalgal compounds have been investigated for their antiproliferative and pro-apoptotic properties. Fucoxanthin, a carotenoid found in brown algae, has been shown to inhibit tumor cell growth by inducing cell-cycle arrest and promoting apoptosis in various cancer-cell lines, including aggressive forms such as triple-negative breast cancer. Its mechanism of action involves suppression of vascular endothelial growth factors VEGF-A and VEGF-C, thereby reducing tumor proliferation and migration (Ahmed et al., 2023). Similarly, astaxanthin has demonstrated synergistic effects with conventional chemotherapeutic agents such as doxorubicin, enhancing cancer-cell apoptosis and reducing viability, which supports its potential role as an adjunct to chemotherapy (Copat et al., 2025).

In metabolic and cardiovascular diseases, clinical evidence supports the efficacy of microalgae in improving key biomarkers. Systematic reviews and meta-analyses have confirmed that *Spirulina* supplementation lowers fasting glucose and HbA1c levels in patients with type 2 diabetes, indicating improved glycemic control (Hamedifard et al., 2019). In parallel, studies in patients with non-alcoholic fatty liver disease (NAFLD) have shown that *Chlorella vulgaris* supplementation leads to significant decreases in serum transaminases (ALT, AST), glucose, and lipid levels, reflecting improvements in liver function and overall metabolic status (Keyghobadi et al., 2024).

Neuroprotective effects have also been documented. In models of Alzheimer's disease, astaxanthin administration mitigated cognitive impairment by reducing oxidative stress via modulation of the SIRT1/PGC-1α signaling axis (Liu et al., 2023a). In ischemia–reperfusion injury models, *Spirulina* supplementation reduced infarct volume and improved neurological outcomes, likely because of its antioxidant effects, highlighting its potential as a neuroprotective agent in cerebrovascular events (Thaakur & Sravanthi, 2010). In the field of infectious diseases, microalgal extracts have shown both antibacterial and antiviral activity. Extracts have inhibited the growth of multiple bacterial pathogens, including methicillin-resistant *Staphylococcus aureus*, suggesting potential as alternative antimicrobial agents in the context of rising antibiotic resistance (Pina-Pérez et al., 2017). In addition, compounds such as calcium spirulan, isolated from *Spirulina platensis*, have demonstrated antiviral activity by preventing the entry of herpes simplex virus type 2 (HSV-2) into host cells, indicating utility in antiviral therapeutic development (Hernández-Corona et al., 2002; Mader et al., 2016).

6 Sustainability Advantages Compared with Current Alternatives

The use of microalgae as a source of therapeutic agents presents substantial sustainability advantages over conventional pharmaceutical and nutraceutical production systems. These advantages span environmental, economic, and social dimensions, supporting the broader goals of sustainable development within the biomedical and biotechnological sectors.

Microalgae cultivation relies on photosynthesis, enabling the conversion of carbon dioxide and inorganic nutrients into biomass using solar energy. In contrast to synthetic drug manufacturing, which typically depends on petroleum-based inputs and energy-intensive processes, microalgal systems can be carbon neutral or even carbon negative when integrated with CO_2 capture technologies, thereby contributing to the reduction of greenhouse-gas emissions (Xia et al., 2021). Their cultivation is also resource efficient. Microalgae can be grown on non-arable land using saline or wastewater sources, which alleviates pressure on freshwater reserves and agricultural land. Moreover, they recycle nutrients from industrial or agricultural effluents, contributing to circular economy strategies, and do not require the application of pesticides or high volumes of synthetic fertilizers (Xia et al., 2021).

In terms of productivity, microalgae outperform many terrestrial crops in biomass yield per unit area. For instance, *Spirulina* can produce more than 20 metric tonnes (t) of dry biomass per hectare annually, whereas crops such as soy yield only a fraction of that. The use of vertical photobioreactors further enhances spatial efficiency and allows for integration into urban and industrial environments (Guidi et al., 2021; Hawrot-Paw & Sąsiadek, 2023).

Environmental impact is also minimized in microalgal systems. Cultivation emits little odor and produces no hazardous waste, and the residual biomass remaining after extraction retains nutritional value. This biomass can be repurposed for animal feed, biogas generation, or agricultural fertilizers, unlike traditional chemical synthesis, which often yields toxic byproducts (Xia et al., 2021).

Microalgal products also offer enhanced safety and purity profiles. Cultures grown under controlled conditions are less likely to accumulate heavy metals or persistent organic pollutants, making them a safer alternative to fish-derived products. For example, oils extracted from microalgae are marketed as free of mercury and dioxins, addressing consumer safety concerns in both food and pharmaceutical applications (Choi & Lee, 2021). From an economic and social perspective, microalgae cultivation promotes green job creation and rural development, particularly in sun-rich, underutilized regions. Its scalability and versatility support decentralized production, offering opportunities for local enterprises and resilience for small-scale producers in developing economies (Di Caprio et al., 2020).

Altogether, microalgae-based biomanufacturing represents a sustainable and adaptable alternative to conventional pharmaceutical supply chains. Although the energy requirements of closed photobioreactor systems remain a challenge, ongoing technological advancements continue to improve efficiency. Life-cycle assessments consistently show that microalgae-derived products such as EPA, DHA, and antioxidant compounds generate fewer greenhouse-gas emissions than fish oil or synthetic production routes (Choi & Lee, 2021).

7 Emerging Technologies and Strategies for Sustainable Production

The efficient production of bioactive molecules from microalgae has advanced significantly thanks to the development of new technologies that optimize cultivation, genetic engineering, and extraction methods, all within a sustainable biorefinery framework.

In terms of cultivation, closed photobioreactors have increasingly replaced open ponds, enabling precise control of parameters such as light, temperature, and CO_2. Studies have shown that specific LED lighting can induce carotenoid accumulation, such as astaxanthin and β-carotene in *Haematococcus pluvialis* and *Dunaliella salina* (Glemser et al., 2016; Liu et al., 2023b). The two-stage cultivation strategy, an initial growth phase followed by a stress phase, has achieved up to 30% increases in carotenoid yields (Cezare-Gomes et al., 2023).

On the genetic front, notable progress has been made using CRISPR-Cas9 genome editing. For example, in *Phaeodactylum tricornutum*, deletion of a repressor gene doubled pigment production (Nymark et al., 2017). Additionally, overexpression of biosynthetic genes has significantly enhanced astaxanthin production in *Haematococcus* and omega-3 lipid accumulation in *Chlorella* (Gu et al., 2024; Norashikin et al., 2018). Heterologous expression has also enabled cyanobacteria to synthesize valuable metabolites from other species (Berla et al., 2013).

Regarding extraction, green methods such as supercritical CO_2 extraction have replaced toxic solvents for lipophilic compounds, producing high-purity, residue-free products (Tzima et al., 2023). For polar compounds, deep eutectic solvents and technologies such as sonication or pulsed electric fields are being explored to achieve efficient cell disruption (Siddhnath et al., 2024).

The biorefinery concept promotes full biomass utilization. For example, after astaxanthin extraction from *Haematococcus*, the residual biomass is reused in aquaculture feeds. In *Spirulina*, sequential extractions of phycocyanin, proteins, and polysaccharides are conducted to maximize economic return (Kaur et al., 2025).

Finally, it is estimated that integrating these strategies has reduced the unit production cost of bioactive compounds by over 50% in the last decade (Bleakley & Hayes, 2017). These advances, along with circular economy principles, position microalgae as an emerging industrial platform for the sustainable production of pharmaceuticals, nutraceuticals, and other therapeutic compounds.

8 Regulatory and Market Challenges

Despite growing optimism surrounding microalgae, their transition from research to clinical and commercial applications faces significant regulatory and market hurdles that must be addressed. These challenges involve ensuring safety and efficacy under regulatory standards, navigating international legal frameworks for approval as drugs or foods, and overcoming economic barriers to widespread market adoption. In the European Union and the USA, products derived from microalgae may fall under different regulatory categories food, novel food, dietary supplement, pharmaceutical, or medical device, depending on their intended use and purity. For instance, several microalgae (such as *Spirulina*, *Chlorella*, *Odontella*) have been approved as novel foods in the EU, which required evidence for toxicological and nutritional safety. In the USA, *Spirulina* and *Chlorella* are GRAS for food use (Karkos et al., 2011), and the FDA has approved specific extracts, e.g., *Haematococcus pluvialis* as a source of astaxanthin in supplements and fortified foods (Su et al., 2023).

There is also a need for clearer guidance on genetically modified (GM) microalgae. For example, transgenic microalgal oils rich in DHA may be evaluated similarly to GM crops in the EU, where labeling regulations are strict, potentially deterring adoption. Yet, if the final product is purified (e.g., 99% astaxanthin), it may be considered chemically identical to its natural counterpart, simplifying the approval process. A successful case is DHA from *Schizochytrium* (a heterotrophic protist related to microalgae), which the FDA approved for use in infant formulas after demonstrating substantial equivalence with fish oil-derived DHA (US Food and Drug Administration, 2017, 2018).

Quality assurance adds another layer of complexity. Variability between production batches and the potential presence of contaminants such as microcystins or heavy metals necessitate strict adherence to Good Manufacturing Practices and ISO standards. Technological advances in photobioreactor design have improved reproducibility, helping to meet these regulatory expectations.

From a clinical standpoint, most evidence supporting microalgal bioactives stems from in vitro or small-scale studies. Demonstrating therapeutic efficacy through robust phase II or III clinical trials remains a significant financial challenge, especially for smaller companies. In the absence of such data, microalgae-based formulations struggle to gain parity with conventional pharmaceutical products.

On the commercial side, production costs continue to be a limiting factor, particularly for compounds that have not yet achieved scale. Despite recent reductions in cost due to technological innovation, building industrial-scale facilities still requires considerable capital investment. Market adoption is further impeded by low consumer awareness, strict limitations on permitted health claims such as those enforced by the European Food Safety Authority and competition from more established alternatives.

In conclusion, advancing the microalgae industry will require regulatory harmonization, stronger clinical validation, improved public awareness, and robust public–private partnerships to foster innovation and industrial scalability. Encouragingly, the increasing global demand for sustainable and natural health solutions offers a promising path forward for the growth of microalgae-based therapeutics.

9 Future Perspectives and Recommendations

Looking ahead, microalgae are emerging as one of the pillars of sustainable biotechnology applied to human health. However, fully realizing their potential still requires targeted research and strategic actions. Below are key perspectives and recommendations to accelerate the transition of microalgae toward higher-impact clinical and commercial applications.

9.1 Exploration of Biodiversity (Bioprospecting)

To date, only a fraction of the tens of thousands of microalgal species has been thoroughly studied. It is highly likely that little-explored species, especially those from extreme environments (thermal, hypersaline, deep-sea), produce novel compounds with unique bioactivities. Bioprospecting programs in diverse ecosystems, coupled with modern metagenomic tools, can reveal promising new strains. For example, Antarctic diatoms could synthesize cryoprotectants useful in medicine, whereas algae from acidic waters might produce unusual antibiotics. Investment in genomic and chemical libraries of microalgae is recommended, along with international collaboration to collect and preserve rare strains. Microalgae banks (e.g., CCMP, SAG) are valuable resources that should be expanded. Computational bioprospecting should also be incorporated using artificial intelligence to predict metabolite production based on genomic pathways, thereby guiding experimental efforts.

9.2 Application of Omics and Systems Biology

To better understand and optimize the production of desired molecules, comprehensive omics approaches—genomics, transcriptomics, and metabolomics—should be applied to microalgae. Building computational models of algal metabolism can help to identify bottlenecks and intervention points. Genomes of several species (e.g., *Chlamydomonas reinhardtii*, *Nannochloropsis gaditana*, *Spirulina*) have been sequenced, but more research is needed to elucidate how metabolic pathways are regulated under varying conditions. Model-assisted metabolic engineering (e.g., genome-scale metabolic models, genetically engineered mouse models) can guide genetic modifications to redirect precursors toward target compounds (Aditi et al., 2025). Next-generation gene editing (high-efficiency CRISPR, base editors, CRISPRa/i) should be implemented across more microalgal species, requiring optimized transformation and cultivation protocols for nonmodel strains. Establishing multidisciplinary consortia integrating algal biologists, metabolic engineers, and data scientists is recommended.

9.3 Large-Scale Clinical Validation

A crucial recommendation is to conduct rigorous clinical trials for the most promising microalgae-based health applications. For example, given the extensive preclinical evidence for *Spirulina* in immunity and inflammation, a multicenter trial in patients with rheumatoid arthritis measuring inflammatory markers and functional improvement could provide the level of evidence needed for official clinical recommendations. The same applies to astaxanthin for neurodegeneration or algal polysaccharides for viral infection prevention. These trials require substantial funding and coordination, ideally from governmental or health organizations as they go beyond the capacity of small enterprises. Without solid clinical data, microalgae will continue to be perceived as alternative dietary supplements. Positive trials, however, could see them included in treatment or prevention guidelines, driving widespread adoption. Therefore, it is recommended that health authorities and funding agencies include microalgae-derived products in their research agendas, treating them on a par with synthetic compounds.

9.4 Policies and Incentives for the Blue Bioeconomy

From a strategic standpoint, governments can play a catalytic role. Incentive policies for the "blue bioeconomy" (marine biotechnology) might include start-up subsidies for cultivation plants, tax breaks for companies developing algal-based drugs, and public–private funding for applied research. Programs such as the European Union's Horizon Europe could allocate funding to demonstrative algal pharmaceutical production projects. Faster regulatory pathways for sustainable innovations (e.g., expedited approval for products with a known safety profile such as *Spirulina*) would be beneficial. Implementing eco-labeling that highlights the low carbon footprint of algal nutraceuticals could provide a competitive advantage. Recommendations include developing national marine biotechnology plans that integrate academia and industry, simplifying administrative procedures for algal installations, and promoting training programs to build an expert workforce in algal cultivation and bioprocessing—integrating microalgae robustly and sustainably into the formal economy.

10 Conclusion

Microalgae represent a highly adaptable and promising biological platform for the generation of bioactive small molecules with substantiated therapeutic properties. Their extensive metabolic and phylogenetic diversity enables the biosynthesis of structurally diverse compounds such as carotenoids, PUFAs, functional peptides, sulfated polysaccharides, and antioxidant polyphenols that are often inaccessible from other natural or synthetic sources. These metabolites exert biological activity through modulation of key pathogenic signaling networks, including NF-κB, Nrf2, PI3K/Akt, and the renin–angiotensin–aldosterone system (RAAS), mediating antioxidant, anti-inflammatory, immunoregulatory, and metabolic benefits.

A growing body of preclinical and clinical research has demonstrated the health-promoting effects of microalgal extracts in various pathological contexts, including chronic inflammation, metabolic dysfunction, and neurodegeneration. Notably, compounds such as astaxanthin and whole-cell preparations of *Spirulina* have shown therapeutic efficacy in animal models and preliminary human studies. Although broader and more rigorous clinical trials remain necessary to validate these findings, the available evidence already supports the biomedical relevance of microalgae-derived interventions.

Beyond their pharmacological profile, microalgae provide a clear ecological advantage over conventional pharmaceutical production systems. Their cultivation leverages photosynthetic carbon fixation, requires minimal land and freshwater resources, and can be seamlessly incorporated into circular bioprocesses using waste-derived nutrients. These attributes align strongly with current global imperatives for sustainable development. As advances in strain engineering, photobioreactor technology, and biorefinery integration continue to reduce production costs and enhance scalability, microalgae are poised to play an increasingly central role in the design of next-generation therapeutics.

References

Adarme-Vega, T. C., Lim, D. K. Y., Timmins, M., Vernen, F., Li, Y., & Schenk, P. M. (2012). Microalgal biofactories: A promising approach towards sustainable omega-3 fatty acid production. *Microbial Cell Factories, 11*(1), 96. https://doi.org/10.1186/1475-2859-11-96

Aditi, B., Yadav, A., Swapnil, P., & Meena, M. (2025). Characterization of microalgal β-carotene and astaxanthin: Exploring their health-promoting properties under the effect of salinity and light intensity. *Biotechnology for Biofuels and Bioproducts, 18*(1), 18. https://doi.org/10.1186/s13068-025-02612-x

Ahmed, S. A., Mendonca, P., Messeha, S. S., & Soliman, K. F. A. (2023). Anticancer effects of Fucoxanthin through cell cycle arrest, apoptosis induction, and angiogenesis inhibition in triple-negative breast cancer cells. *Molecules, 28*(18), 6536. https://doi.org/10.3390/molecules28186536

Akao, Y., Ebihara, T., Masuda, H., Saeki, Y., Akazawa, T., Hazeki, K., Hazeki, O., Matsumoto, M., & Seya, T. (2009). Enhancement of antitumor natural killer cell activation by orally administered spirulina extract in mice. *Cancer Science, 100*(8), 1494–1501. https://doi.org/10.1111/j.1349-7006.2009.01188.x

Ali, E. A. I., Barakat, B. M., & Hassan, R. (2015). Antioxidant and Angiostatic effect of Spirulina platensis suspension in complete Freund's adjuvant-induced arthritis in rats. *PLoS One, 10*(4), e0121523. https://doi.org/10.1371/journal.pone.0121523

Álvarez-Gómez, F., Korbee, N., Casas-Arrojo, V., Abdala-Díaz, R. T., & Figueroa, F. L. (2019). UV Photoprotection, cytotoxicity and immunology capacity of red algae extracts. *Molecules, 24*(2), 341. https://doi.org/10.3390/molecules24020341

Ambati, R., Phang, S.-M., Ravi, S., & Aswathanarayana, R. (2014). Astaxanthin: Sources, extraction, stability, biological activities and its commercial applications – A review. *Marine Drugs, 12*(1), 128–152. https://doi.org/10.3390/md12010128

Ayub, A., Rahayu, F., Khamidah, A., Antarlina, S. S., Iswari, K., Supriyadi, K., Mufidah, E., Singh, A., Chopra, C., & Wani, A. K. (2025). Harnessing microalgae as a bioresource for nutraceuticals: Advancing bioactive compound exploration and shaping the future of health and functional food innovation. *Discover Applied Sciences, 7*(5), 389. https://doi.org/10.1007/s42452-025-06916-3

Bacova, R., Kolackova, M., Klejdus, B., Adam, V., & Huska, D. (2020). Epigenetic mechanisms leading to genetic flexibility during abiotic stress responses in microalgae: A review. *Algal Research, 50*, 101999. https://doi.org/10.1016/j.algal.2020.101999

Baeza-Morales, A., Medina-García, M., Martínez-Peinado, P., Pascual-García, S., Pujalte-Satorre, C., López-Jaén, A. B., Martínez-Espinosa, R. M., & Sempere-Ortells, J. M. (2024). The antitumour mechanisms of carotenoids: A comprehensive review. *Antioxidants, 13*(9), 1060. https://doi.org/10.3390/antiox13091060

Berla, B. M., Saha, R., Immethun, C. M., Maranas, C. D., Moon, T. S., & Pakrasi, H. B. (2013). Synthetic biology of cyanobacteria: Unique challenges and opportunities. *Frontiers in Microbiology, 4*. https://doi.org/10.3389/fmicb.2013.00246

Bhuyar, P., Rahim, M. H., et al. (2020). Exploration of bioactive compounds and antibacterial activity of marine blue-green microalgae (Oscillatoria sp.) isolated from coastal region of West Malaysia. *SN Applied Sciences, 2*(11). https://doi.org/10.1007/s42452-020-03698-8

Bito, T., Okumura, E., Fujishima, M., & Watanabe, F. (2020). Potential of chlorella as a dietary supplement to promote human health. *Nutrients, 12*(9), 2524. https://doi.org/10.3390/nu12092524

Bjørklund, G., Gasmi, A., Lenchyk, L., Shanaida, M., Zafar, S., Mujawdiya, P. K., Lysiuk, R., Antonyak, H., Noor, S., Akram, M., Smetanina, K., Piscopo, S., Upyr, T., & Peana, M. (2022). The role of Astaxanthin as a nutraceutical in health and age-related conditions. *Molecules, 27*(21), 7167. https://doi.org/10.3390/molecules27217167

Bleakley, S., & Hayes, M. (2017). Algal proteins: Extraction, application, and challenges concerning production. *Food, 6*(5), 33. https://doi.org/10.3390/foods6050033

Bortolini, D. G., Maciel, G. M., et al. (2022). Functional properties of bioactive compounds from spirulina spp.: Current status and future trends. *Food Chemistry: Molecular Sciences, 5*, 100134. https://doi.org/10.1016/j.fochms.2022.100134

Cao, J., Wang, J., Wang, S., & Xu, X. (2016). *Porphyra* species: A mini-review of its pharmacological and nutritional properties. *Journal of Medicinal Food, 19*(2), 111–119. https://doi.org/10.1089/jmf.2015.3426

Caroppo, C., & Pagliara, P. (2022). Microalgae: A promising future. *Microorganisms, 10*(8), 1488. https://doi.org/10.3390/microorganisms10081488

Carrizzo, A., Conte, G. M., Sommella, E., Damato, A., Ambrosio, M., Sala, M., Scala, M. C., Aquino, R. P., De Lucia, M., Madonna, M., Sansone, F., Ostacolo, C., Capunzo, M., Migliarino, S., Sciarretta, S., Frati, G., Campiglia, P., & Vecchione, C. (2019). Novel potent decameric peptide of *Spirulina platensis* reduces blood pressure levels through a PI3K/AKT/eNOS-dependent mechanism. *Hypertension, 73*(2), 449–457. https://doi.org/10.1161/HYPERTENSIONAHA.118.11801

Cezare-Gomes, E. A., Lousada, M. E. G., Matsudo, M. C., Ferreira-Camargo, L. S., Ishii, M., Singh, A. K., & Carvalho, J. C. M. (2023). Two-stage semi-continuous cultivation of Dunaliella salina for β-carotene production. *Brazilian Journal of Chemical Engineering, 40*(2), 367–378. https://doi.org/10.1007/s43153-022-00246-2

Cheon, J., Kim, E.-A., Kim, K., Priyadarshana, D. G. C. E., Kang, N., Kim, T., Heo, S.-J., & Cha, S.-H. (2023). Carotenoid Derived from Microalgae Tetraselmis sp. Attenuates LPS-Induced Inflammation through MAPKs/I-kB/NF-kB Signaling Pathway. https://doi.org/10.20944/preprints202310.1383.v1.

Chernova, E., Sidelev, S., Russkikh, I., Voyakina, E., Babanazarova, O., Romanov, R., Kotovshchikov, A., & Mazur-Marzec, H. (2017). Dolichospermum and Aphanizomenon as neurotoxins producers in some Russian freshwaters. *Toxicon, 130*, 47–55. https://doi.org/10.1016/j.toxicon.2017.02.016

Choi, S.-C., & Lee, I.-A. (2021). Effect of MMP/TIMP balancing of Cynoglossus semilaevis Shell extracts on skin protection. *Fishes, 6*(3), 34. https://doi.org/10.3390/fishes6030034

Choo, W.-T., Teoh, M.-L., Phang, S.-M., Convey, P., Yap, W.-H., Goh, B.-H., & Beardall, J. (2020). Microalgae as potential anti-inflammatory natural product against human inflammatory skin diseases. *Frontiers in Pharmacology, 11*. https://doi.org/10.3389/fphar.2020.01086

Copat, C., Favara, C., Tomasello, M., Sica, C., Grasso, A., Dominguez, H., Conti, G., & Ferrante, M. (2025). Astaxanthin in cancer therapy and prevention (review). *Biomedical Reports, 22*(4), 66. https://doi.org/10.3892/br.2025.1944

Crupi, R., Marino, A., & Cuzzocrea, S. (2013). N-3 fatty acids: Role in neurogenesis and neuroplasticity. *Current Medicinal Chemistry, 20*(24), 2953–2963. https://doi.org/10.2174/09298673113209990140

Dang, Y., Li, Z., & Yu, F. (2024). Recent advances in Astaxanthin as an antioxidant in food applications. *Antioxidants, 13*(7), 879. https://doi.org/10.3390/antiox13070879

de Brito, A. F., Silva, A. S., de Oliveira, C. V. C., de Souza, A. A., Ferreira, P. B., de Souza, I. L. L., da Cunha Araujo, L. C., da Silva Félix, G., de Souza Sampaio, R., Tavares, R. L., de Andrade Pereira, R., Neto, M. M., & da Silva, B. A. (2020). Spirulina platensis prevents oxidative stress and inflammation promoted by strength training in rats: Dose-response relation study. *Scientific Reports, 10*(1), 6382. https://doi.org/10.1038/s41598-020-63272-5

Di Caprio, F., Altimari, P., & Pagnanelli, F. (2020). *New strategies enhancing feasibility of microalgal cultivations* (pp. 287–316). https://doi.org/10.1016/B978-0-444-64337-7.00016-1.

Dolganyuk, V., Belova, D., Babich, O., Prosekov, A., Ivanova, S., Katserov, D., Patyukov, N., & Sukhikh, S. (2020). Microalgae: A promising source of valuable bioproducts. *Biomolecules, 10*(8), 1153. https://doi.org/10.3390/biom10081153

Efimova, S. S., Zakharova, A. A., & Ostroumova, O. S. (2020). Alkaloids modulate the functioning of ion channels produced by antimicrobial agents via an influence on the lipid host. *Frontiers in Cell and Developmental Biology, 8*. https://doi.org/10.3389/fcell.2020.00537

Farheen, F., Shreya, K., Mathur, R., & Jha, A. (2024). Natural compounds as epigenetic modulators: Reversing DNA hypermethylation in cancer (review). *World Academy of Sciences Journal, 7*(2), 20. https://doi.org/10.3892/wasj.2024.308

Ferdous, U. T., & Yusof, Z. N. B. (2021). Medicinal prospects of antioxidants from algal sources in cancer therapy. *Frontiers in Pharmacology, 12*. https://doi.org/10.3389/fphar.2021.593116

Ferdous, U. T., Nurdin, A., Ismail, S., Shaari, K., & Yusof, Z. N. B. (2025). A comparative study on antioxidant properties, total phenolics, total flavonoid contents, and cytotoxic properties of marine green microalgae and diatoms. *Journal of Genetic Engineering and Biotechnology, 23*(1), 100456. https://doi.org/10.1016/j.jgeb.2024.100456

Fernandes, R., Campos, J., Serra, M., Fidalgo, J., Almeida, H., Casas, A., Toubarro, D., & Barros, A. I. R. N. A. (2023). Exploring the benefits of Phycocyanin: From spirulina cultivation to its widespread applications. *Pharmaceuticals, 16*(4), 592. https://doi.org/10.3390/ph16040592

Fernández-Rojas, B., Hernández-Juárez, J., & Pedraza-Chaverri, J. (2025). Phycocyanin a nutraceutical with antioxidant activity. In *Improving health and nutrition through bioactive compounds* (pp. 67–78). Elsevier. https://doi.org/10.1016/B978-0-443-21873-6.00004-X

Fernando, I. P. S., Nah, J.-W., & Jeon, Y.-J. (2016). Potential anti-inflammatory natural products from marine algae. *Environmental Toxicology and Pharmacology, 48*, 22–30. https://doi.org/10.1016/j.etap.2016.09.023

Ferrari, M., Muto, A., Bruno, L., & Cozza, R. (2023). DNA methylation in algae and its impact on abiotic stress responses. *Plants, 12*(2), 241. https://doi.org/10.3390/plants12020241

Gallego, R., Valdés, A., Suárez-Montenegro, Z. J., Sánchez-Martínez, J. D., Cifuentes, A., Ibáñez, E., & Herrero, M. (2022). Anti-inflammatory and neuroprotective evaluation of diverse microalgae extracts enriched in carotenoids. *Algal Research, 67*, 102830. https://doi.org/10.1016/j.algal.2022.102830

Ganuza, E., Etomi, E. H., Olson, M., & Whisner, C. M. (2024). Omega-3 eicosapentaenoic polar-lipid rich extract from microalgae Nannochloropsis decreases plasma triglycerides and cholesterol in a real-world normolipidemic supplement consumer population. *Frontiers in Nutrition, 11*. https://doi.org/10.3389/fnut.2024.1293909

Garbuzova-Davis, S., Bickford, C., & P. (2010). Short communication: Neuroprotective effect of spirulina in a mouse model of ALS. *The Open Tissue Engineering and Regenerative Medicine Journal, 3*(1), 36–41. https://doi.org/10.2174/1875043501003010036

Gholipour-Varnami, K., Mohamadnia, S., Tavakoli, O., & Faramarzi, M. A. (2025). A review on the biological activities of key carotenoids: Structures, sources, market, economical features, and stability. *Food Bioscience, 68*, 106529. https://doi.org/10.1016/j.fbio.2025.106529

Ghosh, S., Sarkar, T., Pati, S., Kari, Z. A., Edinur, H. A., & Chakraborty, R. (2022). Novel bioactive compounds from marine sources as a tool for functional food development. *Frontiers in Marine Science, 9*. https://doi.org/10.3389/fmars.2022.832957

Gibney, E. R., & Nolan, C. M. (2010). Epigenetics and gene expression. *Heredity, 105*(1), 4–13. https://doi.org/10.1038/hdy.2010.54

Glemser, M., Heining, M., Schmidt, J., Becker, A., Garbe, D., Buchholz, R., & Brück, T. (2016). Application of light-emitting diodes (LEDs) in cultivation of phototrophic microalgae: Current state and perspectives. *Applied Microbiology and Biotechnology, 100*(3), 1077–1088. https://doi.org/10.1007/s00253-015-7144-6

Gotteland, M., Riveros, K., Gasaly, N., Carcamo, C., Magne, F., Liabeuf, G., Beattie, A., & Rosenfeld, S. (2020). The pros and cons of using algal polysaccharides as prebiotics. *Frontiers in Nutrition, 7*. https://doi.org/10.3389/fnut.2020.00163

Gu, Y., Teo, M. Y. M., In, L. L. A., Shimizu, K., Chae, K.-J., et al. (2024). Genetic engineering of Haematococcus pluvialis microalgae for the enhancement of astaxanthin production: A review. *Biocatalysis and Agricultural Biotechnology, 60*, 103298. https://doi.org/10.1016/j.bcab.2024.103298

Guidi, F., Gojkovic, Z., Venuleo, M., Assunçao, P. A. C. J., & Portillo, E. (2021). Long-term cultivation of a native Arthrospira platensis (spirulina) strain in Pozo Izquierdo (gran Canaria, Spain): Technical evidence for a viable production of food-grade biomass. *PRO, 9*(8), 1333. https://doi.org/10.3390/pr9081333

Gutiérrez, M., Suyama, T. L., Engene, N., Wingerd, J. S., Matainaho, T., & Gerwick, W. H. (2008). Apratoxin D, a potent cytotoxic Cyclodepsipeptide from Papua New Guinea collections of the marine cyanobacteria *Lyngbya majuscula* and *Lyngbya sordida*. *Journal of Natural Products, 71*(6), 1099–1103. https://doi.org/10.1021/np800121a

Hamedifard, Z., Milajerdi, A., Reiner, Ž., Taghizadeh, M., Kolahdooz, F., & Asemi, Z. (2019). The effects of spirulina on glycemic control and serum lipoproteins in patients with metabolic syndrome and related disorders: A systematic review and meta-analysis of randomized controlled trials. *Phytotherapy Research, 33*(10), 2609–2621. https://doi.org/10.1002/ptr.6441

Hawrot-Paw, M., & Sąsiadek, M. (2023). Optimization of microalgal biomass production in vertical tubular photobioreactors. *Energies, 16*(5), 2429. https://doi.org/10.3390/en16052429

Heo, S.-Y., Ko, S.-C., Kim, C. S., Oh, G.-W., Ryu, B., Qian, Z.-J., Kim, G., Park, W. S., Choi, I.-W., Phan, T. T. V., Heo, S.-J., Kang, D.-H., Yi, M., & Jung, W.-K. (2017). A heptameric peptide purified from spirulina sp. gastrointestinal hydrolysate inhibits angiotensin I-converting enzyme- and angiotensin II-induced vascular dysfunction in human endothelial cells. *International Journal of Molecular Medicine, 39*(5), 1072–1082. https://doi.org/10.3892/ijmm.2017.2941

Hernández-Corona, A., Nieves, I., Meckes, M., Chamborro, G., & Barron, B. (2002). Antiviral activity of Spirulina maxima against herpes simplex virus type 2. *Antiviral Research, 56*(3), 279–285. https://doi.org/10.1016/S0166-3542(02)00132-8

Hu, J., Meng, W., Su, Y., Qian, C., & Fu, W. (2023). Emerging technologies for advancing microalgal photosynthesis and metabolism toward sustainable production. *Frontiers in Marine Science, 10*. https://doi.org/10.3389/fmars.2023.1260709

Ighalo, J. O., Dulta, K., Kurniawan, S. B., Omoarukhe, F. O., Ewuzie, U., Eshiemogie, S. O., Ojo, A. U., & Abdullah, S. R. S. (2022). Progress in microalgae application for CO2 sequestration. *Cleaner Chemical Engineering, 3*, 100044. https://doi.org/10.1016/j.clce.2022.100044

Jantaro, S., & Kanwal, S. (2017). Low-molecular-weight nitrogenous compounds (GABA and polyamines) in blue–green algae. In *Algal green chemistry* (pp. 149–169). Elsevier. https://doi.org/10.1016/B978-0-444-63784-0.00008-4

Jiang, L., Wang, Y., Yin, Q., Liu, G., Liu, H., Huang, Y., & Li, B. (2017). Phycocyanin: A potential drug for cancer treatment. *Journal of Cancer, 8*(17), 3416–3429. https://doi.org/10.7150/jca.21058

Kanjan, P., Sahasrabudhe, N. M., de Haan, B. J., & de Vos, P. (2017). Immune effects of β-glucan are determined by combined effects on Dectin-1, TLR2, 4 and 5. *Journal of Functional Foods, 37*, 433–440. https://doi.org/10.1016/j.jff.2017.07.061

Karkos, P. D., Leong, S. C., Karkos, C. D., Sivaji, N., & Assimakopoulos, D. A. (2011). *Spirulina* in clinical practice: Evidence-based human applications. *Evidence-based Complementary and Alternative Medicine, 2011*(1). https://doi.org/10.1093/ecam/nen058

Karpiński, T. M., Ożarowski, M., Alam, R., Łochyńska, M., & Stasiewicz, M. (2021). What do we know about antimicrobial activity of Astaxanthin and Fucoxanthin? *Marine Drugs, 20*(1), 36. https://doi.org/10.3390/md20010036

Karrar, E., Albakry, Z., Mohamed Ahmed, I. A., Zhang, L., Chen, C., Wu, D., & Li, J. (2024). Docosahexaenoic acid and eicosapentaenoic acid from microalgae: Extraction, purification, separation, and analytical methods. *Algal Research, 77*, 103365. https://doi.org/10.1016/j.algal.2023.103365

Kaur, M., Peshwani, H., & Goel, M. (2025). High-value products from microalgae – Unlocking the potential for large-scale commercialization. In *Microalgal biofuels* (pp. 257–276). Elsevier. https://doi.org/10.1016/B978-0-443-24110-9.00012-8

Keyghobadi, H., et al. (2024). Therapeutic potential of lactobacillus casei and Chlorella vulgaris in high-fat diet-induced non-alcoholic fatty liver disease (NAFLD)-associated kidney damages: A stereological study. *Molecular Biology Reports, 51*(1), 613. https://doi.org/10.1007/s11033-024-09542-1

Kolackova, M., Janova, A., Dobesova, M., Zvalova, M., Chaloupsky, P., Krystofova, O., Adam, V., & Huska, D. (2023). Role of secondary metabolites in distressed microalgae. *Environmental Research, 224*, 115392. https://doi.org/10.1016/j.envres.2023.115392

Koo, S. Y., Hwang, J.-H., Yang, S.-H., Um, J.-I., Hong, K. W., Kang, K., Pan, C.-H., Hwang, K. T., & Kim, S. M. (2019). Anti-obesity effect of standardized extract of microalga Phaeodactylum tricornutum containing Fucoxanthin. *Marine Drugs, 17*(5), 311. https://doi.org/10.3390/md17050311

Leão, P. N., Pereira, A. R., Liu, W.-T., Ng, J., Pevzner, P. A., Dorrestein, P. C., König, G. M., Vasconcelos, V. M., & Gerwick, W. H. (2010). Synergistic allelochemicals from a freshwater cyanobacterium. *Proceedings of the National Academy of Sciences, 107*(25), 11183–11188. https://doi.org/10.1073/pnas.0914343107

Lee, A.-H., Shin, H.-Y., Park, J.-H., Koo, S. Y., Kim, S. M., & Yang, S.-H. (2021). Fucoxanthin from microalgae Phaeodactylum tricornutum inhibits pro-inflammatory cytokines by regulating both NF-κB and NLRP3 inflammasome activation. *Scientific Reports, 11*(1), 543. https://doi.org/10.1038/s41598-020-80748-6

Li, Y., Aiello, G., Fassi, E. M. A., Boschin, G., Bartolomei, M., Bollati, C., Roda, G., Arnoldi, A., Grazioso, G., & Lammi, C. (2021). Investigation of Chlorella pyrenoidosa protein as a source of novel angiotensin I-converting enzyme (ACE) and dipeptidyl peptidase-IV (DPP-IV) inhibitory peptides. *Nutrients, 13*(5), 1624. https://doi.org/10.3390/nu13051624

Liu, L., & Rein, K. S. (2010). New peptides isolated from Lyngbya species: A review. *Marine Drugs, 8*(6), 1817–1837. https://doi.org/10.3390/md8061817

Liu, N., Lyu, X., Zhang, X., Zhang, F., Chen, Y., & Li, G. (2023a). Astaxanthin attenuates cognitive deficits in Alzheimer's disease models by reducing oxidative stress via the SIRT1/PGC-1α signaling pathway. *Cell & Bioscience, 13*(1), 173. https://doi.org/10.1186/s13578-023-01129-w

Liu, W., Ji, Y., Long, Y., Huang, W., Zhang, C., Wang, H., Xu, Y., Lei, Z., Huang, W., & Liu, D. (2023b). The role of light wavelengths in regulating algal-bacterial granules formation, protein and lipid accumulation, and microbial functions. *Journal of Environmental Management, 337*, 117750. https://doi.org/10.1016/j.jenvman.2023.117750

Luesch, H., Yoshida, W. Y., Moore, R. E., Paul, V. J., & Corbett, T. H. (2001). Total structure determination of Apratoxin a, a potent novel cytotoxin from the marine cyanobacterium *Lyngbya m ajuscula. Journal of the American Chemical Society, 123*(23), 5418–5423. https://doi.org/10.1021/ja010453j

Mader, J., Gallo, A., Schommartz, T., Handke, W., Nagel, C.-H., Günther, P., Brune, W., & Reich, K. (2016). Calcium spirulan derived from Spirulina platensis inhibits herpes simplex virus 1 attachment to human keratinocytes and protects against herpes labialis. *Journal of Allergy and Clinical Immunology, 137*(1), 197–203.e3. https://doi.org/10.1016/j.jaci.2015.07.027

Maeda, H. (2015). Nutraceutical effects of Fucoxanthin for obesity and diabetes therapy: A review. *Journal of Oleo Science, 64*(2), 125–132. https://doi.org/10.5650/jos.ess14226

Marles, R. J., Barrett, M. L., Barnes, J., Chavez, M. L., Gardiner, P., Ko, R., Mahady, G. B., Dog, T. L., Sarma, N. D., Giancaspro, G. I., Sharaf, M., & Griffiths, J. (2011). United States Pharmacopeia safety evaluation of spirulina. *Critical Reviews in Food Science and Nutrition, 51*(7), 593–604. https://doi.org/10.1080/10408391003721719

Martínez-Sámano, J., Torres-Montes de Oca, A., Luqueño-Bocardo, O. I., Torres-Durán, P. V., & Juárez-Oropeza, M. A. (2018). Spirulina maxima decreases endothelial damage and oxidative stress indicators in patients with systemic arterial hypertension: Results from exploratory controlled clinical trial. *Marine Drugs, 16*(12), 496. https://doi.org/10.3390/md16120496

Masoumi, S., Zokaei, M., Ahmadvand, A., Ghalamkarpour, N., Asadimanesh, N., Azarimatin, A., Fattah, K., Payandeh, Z., Rostami, M., Farjami, A., Pandey, R. K., & Nabi-Afjadi, M. (2025). Microalgae: A treasure trove of anticancer nutraceuticals and promising therapeutic mechanisms. *Advanced Biomedical Research, 14*(1). https://doi.org/10.4103/abr.abr_101_24

Matin, M., Koszarska, M., Atanasov, A. G., Król-Szmajda, K., Jóźwik, A., Stelmasiak, A., & Hejna, M. (2024). Bioactive potential of algae and algae-derived compounds: Focus on anti-inflammatory, antimicrobial, and antioxidant effects. *Molecules, 29*(19), 4695. https://doi.org/10.3390/molecules29194695

Melloni, M., Sergi, D., Simioni, C., Passaro, A., & Neri, L. M. (2023). Microalgae as a nutraceutical tool to antagonize the impairment of redox status induced by SNPs: Implications on insulin resistance. *Biology, 12*(3), 449. https://doi.org/10.3390/biology12030449

Mhadhebi, L., Mhadhebi, A., Robert, J., & Bouraoui, A. (2014). Antioxidant, anti-inflammatory and antiproliferative effects of aqueous extracts of three Mediterranean Brown seaweeds of the genus Cystoseira. *Iranian Journal of Pharmaceutical Research: IJPR, 13*(1), 207–220.

Minois, N. (2014). Molecular basis of the 'anti-aging' effect of spermidine and other natural polyamines – A mini-review. *Gerontology, 60*(4), 319–326. https://doi.org/10.1159/000356748

Moradi, S., Bagheri, R., Amirian, P., Zarpoosh, M., Cheraghloo, N., Wong, A., Zobeiri, M., & Entezari, M. H. (2024). Effects of spirulina supplementation in patients with ulcerative colitis: A double-blind, placebo-controlled randomized trial. *BMC Complementary Medicine and Therapies, 24*(1), 109. https://doi.org/10.1186/s12906-024-04400-w

Mumu, M., Das, A., Emran, T. B., Mitra, S., Islam, F., Roy, A., Karim, M. M., Das, R., Park, M. N., Chandran, D., Sharma, R., Khandaker, M. U., Idris, A. M., & Kim, B. (2022). Fucoxanthin: A promising phytochemical on diverse pharmacological targets. *Frontiers in Pharmacology, 13*. https://doi.org/10.3389/fphar.2022.929442

Münzel, T., & Daiber, A. (2019). Antihypertensive Decapeptide from cyanobacterium *Spirulina platensis*. *Hypertension, 73*(2), 291–293. https://doi.org/10.1161/HYPERTENSIONAHA.118.11893

Nabil-Adam, A., Ashour, M. L., & Shreadah, M. A. (2023). Modulation of MAPK/NF-κB pathway and NLRP3 Inflammasome by secondary metabolites from red algae: A mechanistic study. *ACS Omega, 8*(41), 37971–37990. https://doi.org/10.1021/acsomega.3c03480

Nagayama, Y., Isoo, N., Nakashima, A., Suzuki, K., Yamano, M., Nariyama, T., Yagame, M., & Matsui, K. (2020). Renoprotective effects of paramylon, a β-1,3-D-Glucan isolated from Euglena gracilis Z in a rodent model of chronic kidney disease. *PLoS One, 15*(8), e0237086. https://doi.org/10.1371/journal.pone.0237086

Naselli, F., Volpes, S., Cardinale, P. S., Micheli, S., Cicio, A., Scoglio, G. D., Chiarelli, R., Zizzo, M. G., Picone, P., Caradonna, F., & Nuzzo, D. (2025). First evidence of epigenetic modulation of human gene methylation by microalga Aphanizomenon flos-aquae (AFA) in inflammation-related pathways in intestinal cells. *Clinical Epigenetics, 17*(1), 44. https://doi.org/10.1186/s13148-025-01849-5

Neupert, J., Gallaher, S. D., Lu, Y., Strenkert, D., Segal, N., Barahimipour, R., Fitz-Gibbon, S. T., Schroda, M., Merchant, S. S., & Bock, R. (2020). An epigenetic gene silencing pathway selectively acting on transgenic DNA in the green alga Chlamydomonas. *Nature Communications, 11*(1), 6269. https://doi.org/10.1038/s41467-020-19983-4

Norashikin, M. N., Loh, S. H., Aziz, A., & Cha, T. S. (2018). Metabolic engineering of fatty acid biosynthesis in Chlorella vulgaris using an endogenous omega-3 fatty acid desaturase gene with its promoter. *Algal Research, 31*, 262–275. https://doi.org/10.1016/j.algal.2018.02.020

Nymark, M., Sharma, A., Hafskjold, M., Sparstad, T., Bones, A., & Winge, P. (2017). CRISPR/Cas9 gene editing in the marine diatom Phaeodactylum tricornutum. *Bio-Protocol, 7*(15). https://doi.org/10.21769/BioProtoc.2442

Obluchinskaya, E. D., Pozharitskaya, O. N., & Shikov, A. N. (2022). In vitro anti-inflammatory activities of Fucoidans from five species of Brown seaweeds. *Marine Drugs, 20*(10), 606. https://doi.org/10.3390/md20100606

Okeke, E. S., Ejeromedoghene, O., Okoye, C. O., Ezeorba, T. P. C., Nyaruaba, R., Ikechukwu, C. K., Oladipo, A., & Orege, J. I. (2022). Microalgae biorefinery: An integrated route for the sustainable production of high-value-added products. *Energy Conversion and Management: X, 16*, 100323. https://doi.org/10.1016/j.ecmx.2022.100323

Olguín, E. J., Sánchez-Galván, G., Arias-Olguín, I. I., Melo, F. J., González-Portela, R. E., Cruz, L., De Philippis, R., & Adessi, A. (2022). Microalgae-based biorefineries: Challenges and future trends to produce carbohydrate enriched biomass, high-added value products and bioactive compounds. *Biology, 11*(8), 1146. https://doi.org/10.3390/biology11081146

Oliyaei, N., Moosavi-Nasab, M., Tanideh, N., & Iraji, A. (2023). Multiple roles of fucoxanthin and astaxanthin against Alzheimer's disease: Their pharmacological potential and therapeutic insights. *Brain Research Bulletin, 193*, 11–21. https://doi.org/10.1016/j.brainresbull.2022.11.018

Onyeaka, H., Miri, T., Obileke, K., Hart, A., Anumudu, C., & Al-Sharify, Z. T. (2021). Minimizing carbon footprint via microalgae as a biological capture. *Carbon Capture Science & Technology, 1*, 100007. https://doi.org/10.1016/j.ccst.2021.100007

Osathanunkul, M., Thanaporn, S., Karapetsi, L., Nteve, G. M., Pratsinakis, E., Stefanidou, E., Lagiotis, G., Avramidou, E., Zorxzobokou, L., Tsintzou, G., Athanasiou, A., Mpelai, S., Constandinidis, C., Pantiora, P., Merino, M., Mullor, J. L., Dobrovic, L., Cerasino, L., Ogawa, T., et al. (2025). Diversity of bioactive compounds in microalgae: Key classes and functional applications. *Marine Drugs, 23*(6), 222. https://doi.org/10.3390/md23060222

Ozório, R. Á., Lopes, R. G., Vieira, F., et al. (2024). Crude polysaccharide extract from the microalga Porphyridium cruentum improved nonspecific immune responses and resistance in Penaeus vannamei exposed to vibrio alginolyticus. *Aquaculture Journal, 4*(3), 104–113. https://doi.org/10.3390/aquacj4030008

Panggabean, J. A., Adiguna, S. P., Rahmawati, S. I., Ahmadi, P., Zainuddin, E. N., Bayu, A., & Putra, M. Y. (2022). Antiviral activities of algal-based sulfated polysaccharides. *Molecules, 27*(4), 1178. https://doi.org/10.3390/molecules27041178

Parmar, P., Kumar, R., Neha, Y., & Srivatsan, V. (2023). Microalgae as next generation plant growth additives: Functions, applications, challenges and circular bioeconomy based solutions. *Frontiers in Plant Science, 14*. https://doi.org/10.3389/fpls.2023.1073546

Patel, A. K., Singhania, R. R., Awasthi, M. K., Varjani, S., Bhatia, S. K., Tsai, M.-L., Hsieh, S.-L., Chen, C.-W., & Dong, C.-D. (2021). Emerging prospects of macro- and microalgae as prebiotic. *Microbial Cell Factories, 20*(1), 112. https://doi.org/10.1186/s12934-021-01601-7

Pereira, L., Cotas, J., & Valado, A. (2024). Antioxidants from microalgae and their potential impact on human well-being. *Exploration of Drug Science*, 292–321. https://doi.org/10.37349/eds.2024.00048

Pina-Pérez, M. C., Rivas, A., Martínez, A., & Rodrigo, D. (2017). Antimicrobial potential of macro and microalgae against pathogenic and spoilage microorganisms in food. *Food Chemistry, 235*, 34–44. https://doi.org/10.1016/j.foodchem.2017.05.033

Proença, C., Lobo, M. C., Rufino, A. T., de Oliveira, J. M. P. F., Freitas, M., Trepiana, J., Portillo, M. P., Carvalho, F., & Fernandes, E. (2025). Evidence of phenolic compounds as modulators of ChREBP and NRF2 in type 2 diabetes: A review of their potential for drug development. *Phytochemistry Reviews*. https://doi.org/10.1007/s11101-025-10075-z

Punchakara, A., Prajapat, G., Bairwa, H. K., Jain, S., & Agrawal, A. (2023). Applications of mycosporine-like amino acids beyond photoprotection. *Applied and Environmental Microbiology, 89*(11). https://doi.org/10.1128/aem.00740-23

Remirez, D., González, R., Merino, N., Rodriguez, S., & Ancheta, O. (2002). Inhibitory effects of spirulina in zymosan-induced arthritis in mice. *Mediators of Inflammation, 11*(2), 75–79. https://doi.org/10.1080/09629350220131917

Riccio, G., & Lauritano, C. (2019). Microalgae with immunomodulatory activities. *Marine Drugs, 18*(1), 2. https://doi.org/10.3390/md18010002

Rojas-Villalta, D., Rojas-Rodríguez, D., Villanueva-Ilama, M., Guillén-Watson, R., Murillo-Vega, F., Gómez-Espinoza, O., & Núñez-Montero, K. (2024). Exploring Extremotolerant and Extremophilic microalgae: New Frontiers in sustainable biotechnological applications. *Biology, 13*(9), 712. https://doi.org/10.3390/biology13090712

Ryan, C., Cao, S., Sekiguchi, M., Haraguchi, A., Murata, A., Nakashima, A., Suzuki, K., & Shibata, S. (2023). Euglena gracilis-derived β-glucan paramylon entrains the peripheral circadian clocks in mice. *Frontiers in Nutrition, 10*. https://doi.org/10.3389/fnut.2023.1113118

Ryckebosch, E., Bruneel, C., Termote-Verhalle, R., Goiris, K., Muylaert, K., & Foubert, I. (2014). Nutritional evaluation of microalgae oils rich in omega-3 long chain polyunsaturated fatty acids as an alternative for fish oil. *Food Chemistry, 160*, 393–400. https://doi.org/10.1016/j.foodchem.2014.03.087

Sagnia, B., Fedeli, D., Casetti, R., Montesano, C., Falcioni, G., & Colizzi, V. (2014). Antioxidant and anti-inflammatory activities of extracts from Cassia alata, Eleusine indica, Eremomastax speciosa, Carica papaya and Polyscias fulva medicinal plants collected in Cameroon. *PLoS One, 9*(8), e103999. https://doi.org/10.1371/journal.pone.0103999

Sajilata, M. G., Singhal, R. S., & Kamat, M. Y. (2008). Supercritical CO2 extraction of γ-linolenic acid (GLA) from Spirulina platensis ARM 740 using response surface methodology. *Journal of Food Engineering, 84*(2), 321–326. https://doi.org/10.1016/j.jfoodeng.2007.05.028

Sandgruber, F., Gielsdorf, A., Baur, A. C., Schenz, B., Müller, S. M., Schwerdtle, T., Stangl, G. I., Griehl, C., Lorkowski, S., & Dawczynski, C. (2021). Variability in macro- and micronutrients of 15 commercially available microalgae powders. *Marine Drugs, 19*(6), 310. https://doi.org/10.3390/md19060310

Santigosa, E., Brambilla, F., & Milanese, L. (2021). Microalgae oil as an effective alternative source of EPA and DHA for gilthead seabream (Sparus aurata) aquaculture. *Animals, 11*(4), 971. https://doi.org/10.3390/ani11040971

Sarıtaş, S., Kalkan, A. E., Yılmaz, K., Gurdal, S., Göksan, T., Witkowska, A. M., Lombardo, M., & Karav, S. (2024). Biological and nutritional applications of microalgae. *Nutrients, 17*(1), 93. https://doi.org/10.3390/nu17010093

Sarker, N. K., & Kaparaju, P. (2024). Microalgal bioeconomy: A green economy approach towards achieving sustainable development goals. *Sustainability, 16*(24), 11218. https://doi.org/10.3390/su162411218

Shanaida, M., Mykhailenko, O., Lysiuk, R., Hudz, N., Balwierz, R., Shulhai, A., Shapovalova, N., Shanaida, V., & Bjørklund, G. (2025). Carotenoids for antiaging: Nutraceutical, pharmaceutical, and cosmeceutical applications. *Pharmaceuticals, 18*(3), 403. https://doi.org/10.3390/ph18030403

Sharma, R., Mondal, A. S., & Trivedi, N. (2023). Anticancer potential of algae-derived metabolites: Recent updates and breakthroughs. *Future Journal of Pharmaceutical Sciences, 9*(1), 44. https://doi.org/10.1186/s43094-023-00492-2

Sher, I., Tzameret, A., Peri-Chen, S., Edelshtain, V., Ioffe, M., Sayer, A., Buzhansky, L., Gazit, E., & Rotenstreich, Y. (2018). Synthetic 9-cis-beta-carotene inhibits photoreceptor degeneration in cultures of eye cups from rpe65rd12 mouse model of retinoid cycle defect. *Scientific Reports, 8*(1), 6130. https://doi.org/10.1038/s41598-018-24439-3

Si, P., & Zhu, C. (2022). Biological and neurological activities of astaxanthin (review). *Molecular Medicine Reports, 26*(4), 300. https://doi.org/10.3892/mmr.2022.12816

Siddhnath, S., et al. (2024). Bioactive compounds from micro-algae and its application in foods: A review. *Discover Food, 4*(1), 27. https://doi.org/10.1007/s44187-024-00096-6

Srimongkol, P., Sangtanoo, P., Songserm, P., Watsuntorn, W., & Karnchanatat, A. (2022). Microalgae-based wastewater treatment for developing economic and environmental sustainability: Current status and future prospects. *Frontiers in Bioengineering and Biotechnology, 10*. https://doi.org/10.3389/fbioe.2022.904046

Steadman, C. R., Banerjee, S., Kunde, Y. A., Sanders, C. K., Marrone, B. L., & Twary, S. N. (2020). Inhibition of DNA methylation in Picochlorum soloecismus alters algae productivity. *Frontiers in Genetics, 11*. https://doi.org/10.3389/fgene.2020.560444

Strasky, Z., Zemankova, L., Nemeckova, I., Rathouska, J., Wong, R. J., Muchova, L., Subhanova, I., Vanikova, J., Vanova, K., Vitek, L., & Nachtigal, P. (2013). Spirulina platensis and phycocyanobilin activate atheroprotective heme oxygenase-1: A possible implication for atherogenesis. *Food & Function, 4*(11), 1586. https://doi.org/10.1039/c3fo60230c

Su, G., Jiao, K., Chang, J., Li, Z., Guo, X., Sun, Y., Zeng, X., Lu, Y., & Lin, L. (2016). Enhancing total fatty acids and arachidonic acid production by the red microalgae Porphyridium purpureum. *Bioresources and Bioprocessing, 3*(1), 33. https://doi.org/10.1186/s40643-016-0110-z

Su, M., Bastiaens, L., Verspreet, J., & Hayes, M. (2023). Applications of microalgae in foods, pharma and feeds and their use as fertilizers and biostimulants: Legislation and regulatory aspects for consideration. *Food, 12*(20), 3878. https://doi.org/10.3390/foods12203878

Suo, Q., Yue, Y., Wang, J., Wu, N., Geng, L., & Zhang, Q. (2022). Isolation, identification and *in vivo* antihypertensive effect of novel angiotensin I-converting enzyme (ACE) inhibitory peptides from *spirulina* protein hydrolysate. *Food & Function, 13*(17), 9108–9118. https://doi.org/10.1039/D2FO01207C

Suo, Q., Wang, J., Wu, N., Geng, L., Zhang, Q., & Yue, Y. (2024). Discovery of a novel nanomolar angiotensin-I converting enzyme inhibitory peptide with unusual binding mechanisms derived from Chlorella pyrenoidosa. *International Journal of Biological Macromolecules, 280*, 135873. https://doi.org/10.1016/j.ijbiomac.2024.135873

Thaakur, S., & Sravanthi, R. (2010). Neuroprotective effect of spirulina in cerebral ischemia–reperfusion injury in rats. *Journal of Neural Transmission, 117*(9), 1083–1091. https://doi.org/10.1007/s00702-010-0440-5

Thoré, E. S. J., Muylaert, K., Bertram, M. G., & Brodin, T. (2023). Microalgae. *Current Biology, 33*(3), R91–R95. https://doi.org/10.1016/j.cub.2022.12.032

Tsukamoto, S., Yamanokuchi, R., Yoshitomi, M., Sato, K., Ikeda, T., Rotinsulu, H., Mangindaan, R. E. P., de Voogd, N. J., van Soest, R. W. M., & Yokosawa, H. (2010). Aaptamine, an alkaloid from the sponge Aaptos suberitoides, functions as a proteasome inhibitor. *Bioorganic & Medicinal Chemistry Letters, 20*(11), 3341–3343. https://doi.org/10.1016/j.bmcl.2010.04.029

Tsvetanova, F., & Yankov, D. (2022). Bioactive compounds from red microalgae with therapeutic and nutritional value. *Microorganisms, 10*(11), 2290. https://doi.org/10.3390/microorganisms10112290

Tzima, S., Georgiopoulou, I., Louli, V., & Magoulas, K. (2023). Recent advances in supercritical CO2 extraction of pigments, lipids and bioactive compounds from microalgae. *Molecules, 28*(3), 1410. https://doi.org/10.3390/molecules28031410

Uma, V. S., Usmani, Z., Sharma, M., Diwan, D., Sharma, M., Guo, M., Tuohy, M. G., Makatsoris, C., Zhao, X., Thakur, V. K., & Gupta, V. K. (2023). Valorisation of algal biomass to value-added metabolites: Emerging trends and opportunities. *Phytochemistry Reviews, 22*(4), 1015–1040. https://doi.org/10.1007/s11101-022-09805-4

US Food and Drug Administration. (2017). *GRAS Notice No. 677: Docosahexaenoic acid oil produced in Schizochytrium sp.*

US Food and Drug Administration. (2018). *GRAS Notice No. 732: Docosahexaenoic acid oil produced in Schizochytrium sp.*

Vignaud, J., Loiseau, C., Hérault, J., Mayer, C., Côme, M., Martin, I., & Ulmann, L. (2023). Microalgae produce antioxidant molecules with potential preventive effects on mitochondrial functions and skeletal muscular oxidative stress. *Antioxidants, 12*(5), 1050. https://doi.org/10.3390/antiox12051050

Vujović, T., Paradžik, T., Babić Brčić, S., & Piva, R. (2025). Unlocking the therapeutic potential of algae-derived compounds in hematological malignancies. *Cancers, 17*(2), 318. https://doi.org/10.3390/cancers17020318

Wada, N., Sakamoto, T., & Matsugo, S. (2013). Multiple roles of photosynthetic and sunscreen pigments in cyanobacteria focusing on the oxidative stress. *Metabolites, 3*(2), 463–483. https://doi.org/10.3390/metabo3020463

Wei, Q., Fu, G., Wang, K., Yang, Q., Zhao, J., Wang, Y., Ji, K., & Song, S. (2022). Advances in research on antiviral activities of sulfated polysaccharides from seaweeds. *Pharmaceuticals, 15*(5), 581. https://doi.org/10.3390/ph15050581

Wolf, L., Cummings, T., Müller, K., Reppke, M., Volkmar, M., & Weuster-Botz, D. (2021). Production of β-carotene with *Dunaliella salina* CCAP19/18 at physically simulated outdoor conditions. *Engineering in Life Sciences, 21*(3–4), 115–125. https://doi.org/10.1002/elsc.202000044

Wu, Y., Gao, K., & Riebesell, U. (2010). CO_2 -induced seawater acidification affects physiological performance of the marine diatom *Phaeodactylum tricornutum. Biogeosciences, 7*(9), 2915–2923. https://doi.org/10.5194/bg-7-2915-2010

Wu, S., Liao, X., Zhu, Z., Huang, R., Chen, M., Huang, A., Zhang, J., Wu, Q., Wang, J., & Ding, Y. (2022). Antioxidant and anti-inflammation effects of dietary phytochemicals: The Nrf2/NF-κB signalling pathway and upstream factors of Nrf2. *Phytochemistry, 204*, 113429. https://doi.org/10.1016/j.phytochem.2022.113429

Xia, D., Qiu, W., Wang, X., & Liu, J. (2021). Recent advancements and future perspectives of microalgae-derived pharmaceuticals. *Marine Drugs, 19*(12), 703. https://doi.org/10.3390/md19120703

Xie, J., Chen, X., Wu, J., Zhang, Y., Zhou, Y., Zhang, L., Tang, Y.-J., & Wei, D. (2018a). Antihypertensive effects, molecular docking study, and isothermal titration calorimetry assay of angiotensin I-converting enzyme inhibitory peptides from *Chlorella vulgaris. Journal of Agricultural and Food Chemistry, 66*(6), 1359–1368. https://doi.org/10.1021/acs.jafc.7b04294

Xie, P., Horio, F., Fujii, I., Zhao, J., Shinohara, M., & Matsukura, M. (2018b). A novel polysaccharide derived from algae extract inhibits cancer progression via JNK, not via the p38ï¿½MAPK signaling pathway. *International Journal of Oncology*. https://doi.org/10.3892/ijo.2018.4297

Yang, X., Rong, K., Fu, S., Yang, Y., Liu, S., Zhang, C., Xu, K., Zhang, K., Zhu, Y., Hao, Y., Zhao, J., & Fu, J. (2025). Engineered Spirulina platensis for treating rheumatoid arthritis and restoring bone homeostasis. *Nature Communications, 16*(1), 4434. https://doi.org/10.1038/s41467-025-59579-4

Yao, J., Kim, H. S., Kim, J. Y., Choi, Y.-E., & Park, J. (2020). Mechanical stress induced astaxanthin accumulation of *H. Pluvialis* on a chip. *Lab on a Chip, 20*(3), 647–654. https://doi.org/10.1039/C9LC01030K

Yasuda, K., Nakashima, A., Murata, A., Suzuki, K., & Adachi, T. (2020). Euglena Gracilis and β-Glucan paramylon induce Ca2+ signaling in intestinal tract epithelial, immune, and neural cells. *Nutrients, 12*(8), 2293. https://doi.org/10.3390/nu12082293

Yoon, S., Bay, B. H., & Matsumoto, K. (2025). Harnessing microalgae as sustainable cell factories for polyamine-based Nanosilica for biomedical applications. *Molecules, 30*(8), 1666. https://doi.org/10.3390/molecules30081666

Yu, B. S., Pyo, S., Lee, J., & Han, K. (2024). Microalgae: A multifaceted catalyst for sustainable solutions in renewable energy, food security, and environmental management. *Microbial Cell Factories, 23*(1), 308. https://doi.org/10.1186/s12934-024-02588-7

Zhang, W., Zhou, X., Zhang, Y., Cheng, P., Ma, R., Cheng, W., & Chu, H. (2018). Enhancing astaxanthin accumulation in Haematococcus pluvialis by coupled light intensity and nitrogen starvation in column photobioreactors. *Journal of Microbiology and Biotechnology, 28*(12), 2019–2028. https://doi.org/10.4014/jmb.1807.07008

Zheng, J., Wang, J., Pan, H., Wu, H., Ren, D., & Lu, J. (2017). Effects of IQP, VEP and Spirulina platensis hydrolysates on the local kidney renin angiotensin system in spontaneously hypertensive rats. *Molecular Medicine Reports, 16*(6), 8485–8492. https://doi.org/10.3892/mmr.2017.7602

Chapter 8
Nanotechnology Applications in Human and Animal Health

Nancy Nallely Espinosa Carranza, Anahí Guadalupe Zamora Mendoza, and Fernando Mario Flores Gutiérrez

1 Introduction

Nanotechnology is an emerging multidisciplinary field that has experienced exponential growth worldwide in recent years. It focuses on the development and application of devices, structures, and systems with novel properties and functions, resulting from their nanometric scale size (10^{-9} m, that is, one-billionth of a meter). At this scale, materials exhibit physical, chemical, and biological behaviors that differ significantly from those observed in their macroscopic form (Szczyglewska et al., 2023). These unique properties have been harnessed by the scientific community to address complex challenges and propose innovative solutions across various sectors, including electronics, pharmaceutical sciences, environmental remediation, agriculture, the food industry, and, notably, human and animal health (Chenthamara et al., 2019).

In this context, the application of nanotechnology to the medical sector has led to the emergence of a specialized field known as nanomedicine, which encompasses the use of nanomaterials for the diagnosis, prevention, and treatment of diseases (Kurul et al., 2025). Advances in this field are revolutionizing drug-delivery systems, where their ability to target active ingredients to specific cells or tissues significantly enhances therapeutic efficacy while reducing side effects (Mashweu & Azov, 2024; Patra et al., 2018). This approach has proven particularly useful for diseases with high global prevalence and public-health impact, such as

N. N. E. Carranza (✉) · A. G. Z. Mendoza
Facultad de Ciencias Biológicas, Universidad Autónoma de Nuevo León, Nuevo León, Mexico
e-mail: nancy.espinosacrr@uanl.edu.mx

F. M. F. Gutiérrez
Facultad de Salud Pública y Nutrición, Universidad Autónoma de Nuevo León, Nuevo León, Mexico

M. S. Heya (ed.), *Natural Alternatives and Nanotechnology Applied to the One Health Approach*, https://doi.org/10.1007/978-3-032-03675-9_8

cardiovascular diseases, cancer, ocular disorders, and central nervous system conditions (Anjum et al., 2021). Furthermore, in the field of tissue engineering, nanomedicine has enabled the creation of structures that mimic the architecture and functionality of natural tissues, allowing the integration of living cells and promoting their growth and differentiation. This has opened up new possibilities for the regeneration of bone, skin, and dental tissues (Bellu et al., 2021; Ma et al., 2024). In diagnostics, nanomaterials are being used to enhance the sensitivity, resolution, and specificity of imaging techniques, enabling timely and accurate disease detection.

Beyond its use in humans, nanotechnology has also begun to be applied in the veterinary field. Nanomaterials have been employed to reduce the frequency of vaccine administration and to facilitate their application in both domestic and wild animals (Ianiski et al., 2022). In reproductive biology, they have helped to improve male fertility by increasing sperm longevity and enabling the selection of the most viable sperm for insemination (Feugang et al., 2019). In animal nutrition, nanomaterials have shown the ability to enhance nutrient absorption thanks to their unique transport and uptake properties, promoting better growth and yielding higher quality, contaminant-free animal products (Gelaye, 2024).

Nevertheless, despite their many benefits, the clinical and commercial implementation of nanotechnology still faces significant challenges related to toxicity, biocompatibility, regulatory compliance, and industrial scalability. Therefore, it is essential to continue conducting clinical studies that support its safety and efficacy in both human and veterinary medicine.

This chapter is aimed at providing a comprehensive overview of the impact of nanotechnology on human and animal health by exploring its principles and current applications. By identifying the advances achieved so far, it seeks to promote the development of innovative strategies that guide future research and strengthen the safe and effective use of nanomaterials in biomedical and veterinary fields.

2 Fundamentals of Nanotechnology in Health

2.1 Classification of Nanomaterials

One of the primary criteria for classifying nanomaterials is the degree of spatial confinement, which enables their grouping as follows:

1. Zero-dimensional (0D) nanomaterials, in which all dimensions fall within the nanoscale range (e.g., nanoparticles, fullerenes, and quantum dots).
2. One-dimensional (1D) nanomaterials, where only one dimension extends beyond the nanoscale (e.g., nanorods, nanowires, and nanotubes).
3. Two-dimensional (2D) nanomaterials, characterized by two dimensions exceeding the nanoscale (e.g., nanolayers, nanofilms, and nanosheets).
4. Three-dimensional (3D) nanomaterials or bulk materials, in which none of the dimensions are confined to the nanoscale, but they may incorporate nanoscale

structural components (e.g. nanocomposites, polycrystalline materials, and dendrimers) (Gómez-Gómez et al., 2023; Ma et al., 2024).

In addition to dimensional criteria, nanomaterials can also be classified based on their chemical composition, allowing for the following categories:

1. Organic nanomaterials: micelles, dendrimers, liposomes, and polymeric nanoparticles.
2. Inorganic nanomaterials: metals: silver (Ag), gold (Au), aluminum (Al), cadmium (Cd), copper (Cu), iron (Fe), zinc (Zn), and lead (Pb); metal oxides: copper oxide (CuO), magnesium aluminum oxide ($MgAl_2O_4$), zinc oxide (ZnO), and iron oxide (Fe_3O_4).
3. Carbon-based nanomaterials: fullerenes, graphene, graphene oxide, carbon nanotubes, and carbon nanofibers.
4. Composite nanomaterials: hybrid systems composed of combinations of nanoparticles with each other or with bulk materials (Gómez-Gómez et al., 2023; Ma et al., 2024).

This classification framework provides a foundational understanding of the broad spectrum of available nanomaterials and facilitates the appropriate selection of those most suitable for biomedical applications.

2.2 *Characterization of Nanomaterials*

The properties of nanomaterials are determined by various physicochemical factors, among which size, shape, and surface charge are particularly relevant. These parameters define their biological behavior and compatibility with living systems, enabling their application in both human and veterinary medicine. Understanding these characteristics is essential for the rational design of nanomaterials intended for therapeutic, diagnostic, or preventive purposes.

2.2.1 Particle Size and Surface Area

One of the most critical parameters is particle size, which is closely related to specific surface area. As material dimensions approach the nanoscale, the surface-to-volume ratio increases proportionally, thereby enhancing interactions with the surrounding environment. This promotes higher reactivity, chemical stability, and functional performance (Khan et al., 2022).

The nanoscale size allows nanomaterials to traverse physiological barriers such as the skin, olfactory mucosa, and the blood–brain barrier. Once inside the body, they can circulate via the bloodstream, accumulate in target organs, and interact with subcellular structures. Therefore, both particle size and surface area are key determinants in the biodistribution, accumulation, and clearance of nanomaterials within the organism (Gatoo et al., 2014).

A clear example of the influence of size in biomedical applications is that of silver nanoparticles (AgNPs). Their nanoscale dimensions enhance penetration through the skin barrier, especially when it is compromised. Moreover, AgNPs exhibit affinity for both microbial and mammalian cells and display potential antimicrobial activity. These characteristics have led to their incorporation into medical devices such as catheters, dressings, implants, and antiseptic solutions, where they contribute to modulating inflammatory responses and preventing infections.

2.2.2 Shape of Nanomaterials

In addition to size, the shape of nanomaterials plays a significant role in determining their biological behavior. These structures may adopt a wide variety of morphologies, including spheres, tubes, fibers, cylinders, spirals, flat sheets, or cones (Joudeh & Linke, 2022). It has been demonstrated that nanomaterials with similar composition and size can exhibit distinct physicochemical and biological properties depending on their structural shape. For instance, spherical nanoparticles typically show higher cellular internalization rates than cubic ones, followed by rod- and disc-shaped particles. Likewise, rod-shaped nanoparticles have a greater contact area with membrane receptors than spherical ones, which reduces the number of available binding sites and influences their biological response (Yagublu et al., 2022).

2.2.3 Surface Charge of Nanomaterials

Surface charge is another critical factor, as it governs the interaction of nanomaterials with other charged entities in their environment. It has been observed that positively charged or neutral nanostructures are more readily taken up by cells than negatively charged ones, owing to electrostatic repulsion between anionic surfaces and the negatively charged cell membrane (Yagublu et al., 2022).

Furthermore, surface charge affects the colloidal stability of the nanomaterial system. Electrostatic repulsion between particles helps to prevent aggregation, thereby promoting uniform behavior in biological media (Szczyglewska et al., 2023).

The physicochemical characterization of nanomaterials enables the prediction and optimization of their performance in biological systems, both human and animal (Fig. 8.1). Owing to their unique properties, nanomaterials have emerged as one of the most promising technological advancements of the twenty-first century. In the following section the most relevant applications of these systems in human and veterinary medicine are discussed, highlighting their therapeutic, diagnostic, and preventive potential.

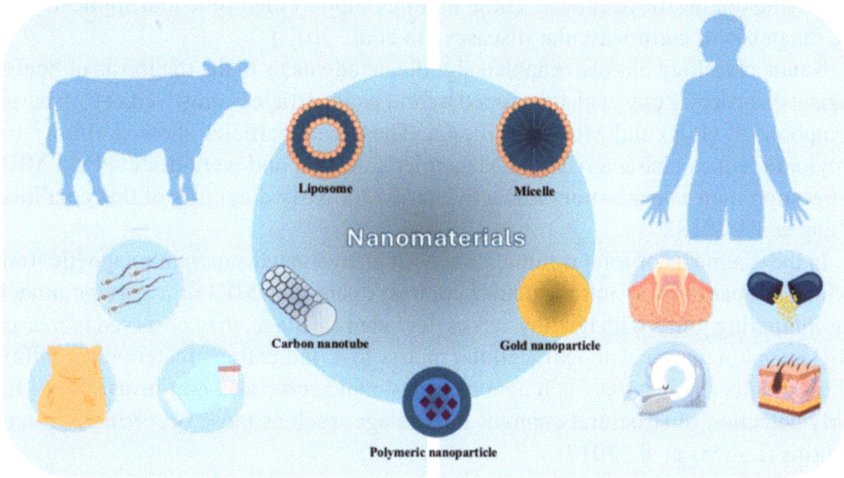

Fig. 8.1 Biomedical applications of nanomaterials in human and veterinary health

3 Applications of Nanotechnology in Human Health

3.1 Nanotechnological Advances in Medical Diagnostic Techniques

Early disease detection is a critical factor for therapeutic success and favorable patient prognosis. Currently, imaging techniques such as X-rays, magnetic resonance imaging (MRI), computed tomography (CT), positron emission tomography (PET), endoscopy, and ultrasound are well-established and widely used in diagnosing various pathologies, including trauma, cardiovascular conditions, and neurological disorders (Hussain et al., 2022). However, these techniques have certain limitations, particularly in detecting pathological signals at early stages, owing to insufficient spatial resolution or low sensitivity. Additionally, conventional contrast agents used in these methods have been associated with adverse effects, limiting their use in certain patients. Other challenges include high equipment costs, complex procedures, and the need for highly specialized personnel, all of which restrict accessibility, especially in developing countries (Liu et al., 2024).

In this context, nanotechnology has emerged as a promising alternative with the potential to revolutionize medical imaging. Among the most notable nanodiagnostic strategies is the development of targeted probes and optimized contrast agents. For example, Hang et al. designed superparamagnetic iron-oxide nanoparticles functionalized with antibodies targeting the GPIIb/IIIa platelet receptor. These nanoparticles acted as dual-contrast agents for MRI, enabling precise visualization of human thrombi in vitro and in murine models of carotid thrombosis, appearing as bright signals on T1-weighted images and dark signals on T2-weighted images, thereby improving diagnostic accuracy. These findings support their potential application in the diagnosis of cardiovascular diseases (Ta et al., 2017).

Nanotechnology has also enabled significant advances in the diagnosis of neurological disorders. Zeng et al. developed hybrid nanoparticles composed of Pittsburgh compound B (PiB) and $Mn_{0.6}$ $Zn_{0.4}Fe_2O_4$. These nanoparticles showed affinity for amyloid plaques characteristic of Alzheimer's disease and were detected via MRI, suggesting their use as a nonradioactive probe for early diagnosis of this condition (Zeng et al., 2018).

In the diagnosis of joint trauma, Labens et al. evaluated superparamagnetic iron-oxide nanoparticles as intra-articular contrast agents for MRI in a porcine model. By comparing joints with healthy versus degraded cartilage, they observed increased MRI signal in tissues with extracellular matrix loss, indicating greater permeability. These results suggest that such nanoparticles could serve as a non-invasive tool for early detection of structural changes in cartilage, such as those occurring in osteoarthritis (Labens et al., 2017).

Together, these developments demonstrate that nanotechnology has significantly impacted the field of medical imaging by enabling earlier, more accurate, and more accessible diagnostics, thus enhancing the performance of currently used clinical technologies.

3.2 *Nanotechnological Advances in Dental Tissues*

Oral health plays a vital role in overall health and quality of life. Among the most common oral diseases are dental caries, periodontal diseases, and, in advanced cases, tooth loss. In this context, nanotechnology has emerged as a promising tool offering innovative, non-invasive solutions for the treatment and prevention of various oral conditions. For example, AgNPs have demonstrated bactericidal effects against *Streptococcus mutans*, a microorganism responsible for biofilm formation on the tooth surface, commonly known as dental plaque. These nanoparticles act by disrupting bacterial cell membranes and inhibiting enzymatic processes, interfering with bacterial viability and colonization capacity (Saliminasab et al., 2022).

In another study, nanometric zinc oxide (ZnO) coatings were applied to nickel-titanium (NiTi) orthodontic wires. The results showed that these nanocoatings not only inhibited plaque formation around the orthodontic appliances but also reduced friction between components, minimizing patient discomfort and irritation (Hammad et al., 2020).

Dental fillings are essential for restoring tooth structure and preventing recurrent caries. In this regard, incorporating zirconia and glass silica nanoparticles into dental resins significantly enhanced their flexural and tensile strength, contributing to greater material durability and reducing crack propagation (Alshamrani et al., 2023).

Collectively, these advances reflect the enormous potential of nanotechnology in the dental field, both for prevention and treatment, through the design of functional materials with antimicrobial and mechanical properties.

3.3 Nanotechnological Advances in Skin Tissues

The skin is the human body's first line of defense, continuously exposed to external stimuli, and plays a crucial role in protecting against injuries and damage. Therefore, skin care and regeneration represent areas of high interest in clinical practice. Nanomaterials have been utilized to maintain and restore cutaneous homeostasis, with particular focus on wound healing and tissue regeneration. In one study, gold and silver nanoclusters were developed to evaluate their healing effects on rat skin wound models in vivo. The results demonstrated that the nanoclusters increased wound healing rates and reduced inflammation at the treatment site (Li et al., 2021).

3.4 Nanotechnological Advances in Targeted Drug Delivery and Natural Bioactive Compounds

Most commonly used drugs do not act selectively on specific organs, tissues, or cells, but rather distribute systemically throughout the body. To reach their site of action, they must cross multiple biological barriers, which can diminish their therapeutic efficacy. During this process, active compounds may accumulate in non-target tissues, potentially causing adverse effects. As a result, high doses are often required to achieve therapeutic concentrations, increasing toxicity and reducing patient adherence to treatment (Table 8.1).

In this context, nanomaterials have proven to be promising tools for improving drug delivery. Their ability to target specific tissues or organs and to release active compounds in a controlled manner not only enhances therapeutic efficacy but also minimizes side effects and improves treatment adherence. A representative example is paclitaxel (Ptx), a hydrophobic taxane with low water solubility that is typically administered in formulations containing ethanol and polyoxyethylated castor oil. These excipients have been associated with hypersensitivity reactions, necessitating premedication with antihistamines and corticosteroids. To overcome these limitations, Ptx bound to albumin was encapsulated in nanoparticles, enabling dissolution in physiological saline without the need for toxic solvents. This strategy not only improved the drug's solubility but also increased the free fraction of the active compound, thereby enhancing its therapeutic effect (Paluszkiewicz et al., 2022).

Furthermore, the combined use of nanotechnology and natural products such as essential oils or plant extracts has garnered increasing interest in recent years. Natural products are particularly attractive owing to their diverse biological activities, including antioxidant, antimicrobial, antiseptic, and antiviral properties (Armendáriz-Barragán et al., 2016; Mancianti & Ebani, 2020). However, their application faces limitations such as high volatility, poor water solubility, and chemical instability when exposed to light, oxygen, or pH changes (Badgujar Vaishnavi et al., 2021). Incorporating these compounds into nanostructured systems has enhanced their bioavailability, stability, controlled release, and therapeutic efficacy.

Table 8.1 Use of nanomaterials in human health

Sector	Type of nanomaterial	Application	References
Medical diagnostic techniques	Hybrid micelles composed of metal oxides and peptides	Contrast agent for MRI of thrombi in atherosclerotic plaques	Poon et al. (2018)
	Manganese carbonate nanoparticles	pH-responsive nanoparticles that enhance MRI for the detection of malignant tumors and tumor metastasis	Zhu et al. (2021)
	Functionalized iron oxide nanoparticles	Hybrid PET–MRI contrast agents for detecting MMP-2 enzyme activity in carotid atherosclerotic plaques in murine models	Tu et al. (2022)
	Magnetic nanoprobe	Improves imaging of hypoxic tumor regions in breast cancer	Ma et al. (2024)
Dental tissues	Gold nanoparticles	Stimulation of periodontal regeneration	Zhang et al. (2021)
	Gold nanoparticles	Effective antimicrobial activity against *Streptococcus oralis*	Al-Fahham et al. (2023)
	Silver nanoparticles	Toothpaste supplemented with nanoparticles exhibiting antimicrobial activity	Ahmed et al. (2023)
Skin tissues	PCL nanofibrous scaffolds with magnetite and silver nanoparticles	Wound dressings	Ahmed et al. (2021)

MRI magnetic resonance imaging, *PCL* poly(ε-caprolactone), *PET* positron emission tomography

For example, Yen et al. developed nanoemulsions based on andrographolide, a diterpenoid with poor water solubility but notable anti-inflammatory and antioxidant properties. The results showed that encapsulation increased its bioavailability sixfold compared with conventional suspensions and improved its efficacy in preventing gastrointestinal inflammatory disorders (Yen et al., 2018).

In summary, these advances demonstrate that nanotechnology provides an innovative and effective approach to targeted delivery of drugs and bioactive compounds by enabling localized action in specific tissues and minimizing undesired side effects.

4 Applications of Nanotechnology in Animal Health

4.1 Reproductive Performance in Livestock

Reproduction is a critical trait in livestock systems, ensuring species continuity and directly impacting production efficiency and costs (Liang et al., 2023). Traditional reproductive management strategies include nutritional control, the use of biostimulatory tools (e.g., male effect), prevention of reproductive diseases, and assisted reproductive technologies such as artificial insemination, estrus synchronization, and embryo transfer (Gulzar et al., 2024). However, the adoption of these methods is often limited by high operational costs, lack of infrastructure, and insufficient technical expertise (Choudhary et al., 2016).

Recent advances in nanotechnology have introduced innovative strategies to enhance reproductive efficiency. One such approach is sperm nanopurification, which utilizes magnetic nanoparticles coated with antibodies to selectively remove defective spermatozoa (Hill & Li, 2017). For instance, Paul et al. (2024) developed lectin-functionalized iron-oxide magnetic nanoparticles capable of eliminating sperm with high acrosome reactivity. A dose of 0.6 mg effectively improved semen quality without compromising mitochondrial membrane potential or increasing reactive oxygen species production.

Another promising application involves hormonal nanodrug delivery systems, which facilitate the administration of reproductive hormones such as steroids or gonadotropins (Hashem & Gonzalez-Bulnes, 2021). For example, Oliveira et al. (2013) engineered polylactic acid nanofiber membranes loaded with progesterone. In vitro release studies revealed a first-order kinetic profile, supporting controlled hormone delivery for estrus regulation. Similarly, Karuppannan et al. (2017) produced electrospun zein nanofibers incorporating 1.2 g of progesterone, demonstrating sustained release behavior suitable for estrus synchronization in cattle.

These hormone-loaded nanopharmaceuticals significantly enhance the biological efficacy of treatments by increasing hormone stability, improving passage through epithelial or endothelial barriers, facilitating systemic absorption, and enabling site-specific release. As a result, lower dosages can be employed without compromising efficacy (Hashem & Gonzalez-Bulnes, 2020).

4.2 Animal Nutrition

Nanotechnology in animal nutrition primarily involves the administration of essential nutrients such as minerals, vitamins, and probiotics (Gelaye, 2024). These can be delivered through the incorporation of both organic (e.g., sugars, fats, proteins) and inorganic nanoparticles (e.g., calcium, silica, silver, magnesium) into feed formulations (Abdelnour et al., 2021). Owing to their unique physicochemical characteristics (particularly their nanoscale size and high surface-to-volume ratio) these nanonutrients offer several advantages:

1. Enhanced bioavailability via improved gastrointestinal absorption
2. Reduced need for preservatives and masking of undesirable odors
3. Improved solubility and dispersibility
4. Protection from environmental degradation factors (e.g., light, oxidation, pH, enzymatic activity, and temperature) (Barwant et al., 2024).

Collectively, these features contribute to improved growth performance, digestive efficiency, immune response, reproductive performance, and gut microbiota homeostasis (Almeida et al., 2024).

Notable examples include the nanoencapsulation of silymarin for broiler chickens, which enhances its antioxidant and hepatoprotective properties (Yousefdoost et al., 2019). ZnO nanoparticles have been shown to improve weight gain in Awassi lambs (Maki et al., 2024). Nano-selenium is particularly well studied; its incorporation into feed improves weight gain, milk production, nutrient intake, and semen quality, as well as increasing pregnancy rates (Sitaresmi et al., 2024).

Essential oils (EOs) have also been evaluated as feed additives owing to their antimicrobial, antioxidant, and digestive stimulant properties (Nehme et al., 2021). However, their high volatility, instability, intense odor, and poor solubility limit direct use in feed (Neculai-Valeanu et al., 2021). Nanotechnology offers a viable solution: nanoencapsulation enhances EO stability and bioavailability, leading to improved feed intake, weight gain, and feed conversion ratios (Linh et al., 2022).

For instance, Mendoza-Ordoñez et al. compared nanoencapsulated blends of soursop, lemon, and eucalyptus EOs with halquinol in broiler diets. Blends T1 (33.33% each EO) and T4 (25% soursop and lemon, 50% eucalyptus) yielded the highest average daily weight gain and breast-muscle production, along with significantly reduced lipid peroxidation. Bromatological analysis further revealed improved crude protein, fat, and ash content in T4-treated birds. Additionally, Adil et al. (2024) demonstrated that 200 mg/kg of nanoencapsulated rosemary EO improved body-weight gain, nutrient digestibility, and meat quality in broilers.

Nanoparticle-based EO supplementation is also being explored in aquaculture. Yousefi et al. (2025) investigated the effects of *Laurus nobilis* EO-loaded nanoliposomes on Nile tilapia (*Oreochromis niloticus*). Treated fish showed enhanced growth and intestinal enzyme activity (amylase, protease, lipase). Moreover, upon experimental infection with *Aeromonas hydrophila*, fish fed 50 or 100 mg/kg of nanoliposomes had significantly higher survival rates. The results support the use of 50 mg/kg EO-nanoliposomes as an effective growth promoter, immune booster, and pathogen-resistance enhancer in tilapia.

4.3 Disease Diagnosis

Veterinary disease diagnosis traditionally relies on techniques such as serological testing, cell cultures, immunoassays, immunodiffusion, electron microscopy, and polymerase chain reaction (Suminda et al., 2022). Despite their utility, these methods often require complex laboratory infrastructure, trained personnel, long processing times, and may exhibit limited sensitivity or specificity, particularly for early-stage detection or low pathogen loads (Zhao et al., 2025).

Consequently, there is growing interest in the development of rapid, portable, and reliable diagnostic tools, notably biosensors. These devices integrate a biological recognition element (e.g., enzymes, antibodies, aptamers, glycans, DNA probes) with a transducer to generate a measurable signal proportional to the concentration of a target analyte (Naresh & Lee, 2021; Rios et al., 2024).

To enhance biosensor performance, nanotechnology has introduced various nanomaterials, including metal nanoparticles (e.g., Ag, Au), metal-oxide nanoparticles (e.g., magnesium oxide, ZnO), magnetic nanoparticles, and dendrimers, which significantly improve sensitivity, specificity, portability, and cost-effectiveness (Mondal et al., 2022; Dar et al., 2020). These so-called nanobiosensors represent a powerful tool for disease surveillance and management in livestock (Table 8.2).

A notable application is in the detection of *Brucella* spp. Abuawad et al. (2023) developed a magnetic nanoparticle-based biosensor capable of selectively detecting hybridized *Brucella* DNA. The sensor exhibited a detection time of only 10 min and a sensitivity threshold of 55 copies, offering a robust, rapid, and field-deployable diagnostic tool for early brucellosis detection and control.

5 Conclusion

Nanotechnology represents a disruptive platform with vast potential to transform human, animal, and environmental health systems. Its capacity to engineer nanomaterials with tailored properties has enabled significant advancements in diagnostics, therapeutics, tissue regeneration, and nutrition. In veterinary applications, these developments contribute not only to improved productivity but also to animal welfare and food safety—key components of the One-Health approach. Furthermore, nanotechnology can support zoonotic disease control, sustainable agricultural practices, and environmental monitoring, highlighting its integrative role across interconnected ecosystems. However, the translation of laboratory breakthroughs into clinical or field applications requires addressing fundamental challenges related to toxicological safety, biological compatibility, regulatory frameworks, and industrial scalability. Promoting interdisciplinary and cross-sectoral research that bridges nanotechnology with medicine, veterinary sciences, pharmacology, engineering, and environmental science is essential to ensure the responsible, equitable, and sustainable development of these technologies in pursuit of global health goals.

Table 8.2 Use of nanomaterials for diagnosis, reproduction, and nutrition in livestock animals

Sector	Nanomaterials	Application	References
Diagnosis of diseases	Sequencing of nanopores	Detection of *Mycoplasma Bovis*	Bokma et al. (2021)
	Gold nanoparticles and multi-walled carbon nanotubes	Detection of toxin 32^{a}-β2 for the identification of *Clostridium perfringens*	Yang et al. (2024)
	Gold nanoparticles	Detection of brucellosis antibodies	Lv et al. (2018)
	Silica nanoparticles	Detection of antigens for the identification of avian influenza	Zou et al. (2012)
	Gold nanoparticles	Detection of aptamer for the identification of *Brucella abortus* and *Brucella melitensis*	Ahangari et al. (2024)
	Zinc-oxide nanoparticles	Detection of brucellosis antibodies	Wahab et al. (2017)
	Magnesium-oxide nanoparticles	Detection of antibodies for the identification of *Brucella abortus*	Khan et al. (2017)
	Porous Au/PtAu nanotubes grafted with cyclodextrin	Detection of antibodies for the identification of porcine reproductive and respiratory syndrome virus	Shao et al. (2017)
	Polydopamine–silver nanoparticles	Detection of RNA for the identification of porcine reproductive and respiratory syndrome virus	Yu et al. (2023)
	Gold nanoparticles	Detection of aptamer for the identification of porcine reproductive and respiratory syndrome virus	Kuitio et al. (2020)
	Gold nanowires	Detection of antibodies for the identification of bovine viral diarrhea	Montrose et al. (2015)
	Gold nanoparticles	Detection of aptamer for the identification of bovine viral diarrhea	Rabiei et al. (2024)
	Nickel hexacyanoferrate nanoparticles and gold nano cages	Detection of antibodies for the identification of *Streptococcus suis*	Zhu et al. (2015)
Nutrition	Selenium nanoparticles	Improve ruminal digestibility and fermentation	Shakweer et al. (2023)
	Copper-oxide nanoparticles	Improving copper oxide bioavailability in the presence of dietary antagonists in dairy cows	Williams et al. (2024)
	Zinc-oxide nanoparticles	Improve performance, ruminal fermentation, and total antioxidant capacity in sheep	Hosseini-Vardanjani et al. (2020)
Reproduction	Lectin-coated magnetic nanoparticles	Nanopurification of donkey semen	Yousef et al. (2020)
	Magnetic nanoparticles	Cryopreservation of ram semen	Moradi et al. (2022)
	Magnetic nanoparticles coated with antibodies against ubiquitin	Nanopurification of bull semen	Odhiambo et al. (2014)

References

Abdelnour, S. A., Alagawany, M., Hashem, N. M., Farag, M. R., Alghamdi, E. S., Hassan, F. U., Bilal, R. M., Elnesr, S. S., Dawood, M. A. O., Nagadi, S. A., Elwan, H. A. M., et al. (2021). Nanominerals: Fabrication methods, benefits and hazards, and their applications in ruminants with special reference to selenium and zinc nanoparticles. *Animals, 11*(7), 1916.

Abuawad, A., Ashhab, Y., Offenhäusser, A., & Krause, H. J. (2023). DNA sensor for the detection of Brucella spp. based on magnetic nanoparticle markers. *International Journal of Molecular Sciences, 24*(24), 17272.

Adil, S., Banday, M. T., Hussain, S. A., Wani, M. A., Al-Olayan, E., Patra, A. K., et al. (2024). Impact of nanoencapsulated rosemary essential oil as a novel feed additive on growth performance, nutrient utilization, carcass traits, meat quality and gene expression of broiler chicken. *Food, 13*(10), 1515.

Ahangari, A., Mahmoodi, P., Zolfigol, M. A., Mohammadzadeh, A., & Salouti, M. (2024). Rapid detection of Brucella cells using a gold nanoparticle-based aptasensor via a simple colorimetric method. *BMC Veterinary Research, 20*(1), 513.

Ahmed, M. K., Zayed, M. A., El-dek, S. I., Hady, M. A., El Sherbiny, D. H., & Uskoković, V. (2021). Nanofibrous ε-polycaprolactone scaffolds containing ag-doped magnetite nanoparticles: Physicochemical characterization and biological testing for wound dressing applications in vitro and in vivo. *Bioactive Materials, 6*(7), 2070–2088. https://doi.org/10.1016/j.bioactmat.2020.12.026

Ahmed, O. A. K., Sibuyi, N. R. S., Fadaka, A. O., Maboza, E., Olivier, A., Madiehe, A. M., et al. (2023). Prospects of using gum Arabic silver nanoparticles in toothpaste to prevent dental caries. *Pharmaceutics, 15*(3). https://doi.org/10.3390/pharmaceutics15030871

Al-Fahham, B.M., Mohamed, R.A., Taqi Al-Talqani, J.M., Fahad, A.H., Haider, J. (2023). Evaluating Antimicrobial Effectiveness of Gold Nanoparticles against Streptococcus oralis. *International Journal of Dentistry*, 1–8. https://doi.org/10.1155/2023/9935556

Almeida, C. F., Faria, M., Carvalho, J., & Pinho, E. (2024). Contribution of nanotechnology to greater efficiency in animal nutrition and production. *Journal of Animal Physiology and Animal Nutrition, 108*(5), 1430–1452.

Alshamrani, A., Alhotan, A., Kelly, E., & Ellakwa, A. (2023). Mechanical and biocompatibility properties of 3D-printed dental resin reinforced with glass silica and zirconia nanoparticles: In vitro study. *Polymers, 15*(11). https://doi.org/10.3390/polym15112523

Anjum, S., Ishaque, S., Fatima, H., Farooq, W., Hano, C., Abbasi, B. H., & Anjum, I. (2021). Emerging applications of nanotechnology in healthcare systems: Grand challenges and perspectives. *Pharmaceuticals, 14*(8), 1–26. https://doi.org/10.3390/ph14080707

Armendáriz-Barragán, B., Zafar, N., Badri, W., Galindo-Rodríguez, S. A., Kabbaj, D., Fessi, H., & Elaissari, A. (2016). Plant extracts: From encapsulation to application. *Expert Opinion on Drug Delivery, 13*(8), 1165–1175. https://doi.org/10.1080/17425247.2016.1182487

Badgujar Vaishnavi, C., Shinde Ananya, J., Nehete Jitendra, Y., & Bhambar Rajendra, S. (2021). Strategies to improve stability of essential oils. *Journal of Pharmaceutical Sciences and Research, 13*(7), 416–425.

Barwant, M. M., Rasool, A., Mannam, V. K., Hussain, A., Abdullah, M., et al. (2024). Nanotechnology applications in livestock farming and animal health. In *Harnessing NanoOmics and Nanozymes for sustainable agriculture* (pp. 413–442). IGI Global.

Bellu, E., Medici, S., Coradduzza, D., Cruciani, S., Amler, E., & Maioli, M. (2021). Nanomaterials in skin regeneration and rejuvenation. *International Journal of Molecular Sciences, 22*(13). https://doi.org/10.3390/ijms22137095

Bokma, J., Vereecke, N., Pas, M. L., Chantillon, L., Vahl, M., Weesendorp, E., et al. (2021). Evaluation of nanopore sequencing as a diagnostic tool for the rapid identification of mycoplasma bovis from individual and pooled respiratory tract samples. *Journal of Clinical Microbiology, 59*(12), 10–1128.

Chenthamara, D., Subramaniam, S., Ramakrishnan, S. G., Krishnaswamy, S., Essa, M. M., Lin, F. H., & Qoronfleh, M. W. (2019). Therapeutic efficacy of nanoparticles and routes of administration. *Biomaterials Research, 23*(1), 1–29. https://doi.org/10.1186/s40824-019-0166-x

Choudhary, K. K., Kavya, K. M., Jerome, A., & Sharma, R. K. (2016). Advances in reproductive biotechnologies. *Veterinary world, 9*(4), 388.

Dar, F. A., Qazi, G., & Pirzadah, T. B. (2020). Nano-biosensors: NextGen diagnostic tools in agriculture. In *Nanobiotechnology in agriculture: An approach towards sustainability* (pp. 129–144). Springer International Publishing.

Feugang, J. M., Rhoads, C. E., Mustapha, P. A., Tardif, S., Parrish, J. J., Willard, S. T., & Ryan, P. L. (2019). Treatment of boar sperm with nanoparticles for improved fertility. *Theriogenology, 137*, 75–81. https://doi.org/10.1016/j.theriogenology.2019.05.040

Gatoo, M. A., Naseem, S., Arfat, M. Y., Mahmood Dar, A., Qasim, K., & Zubair, S. (2014). Physicochemical properties of nanomaterials: Implication in associated toxic manifestations. *BioMed Research International, 2014*. https://doi.org/10.1155/2014/498420

Gelaye, Y. (2024). Application of nanotechnology in animal nutrition: Bibliographic review. *Cogent Food and Agriculture, 10*(1). https://doi.org/10.1080/23311932.2023.2290308

Gómez-Gómez, A. L., Martínez-Ayala, A. L., et al. (2023). Relationship of nanomaterials' structure based on their application in the food industry: Physicochemical and techno-functional characteristics. *Applied Sciences (Switzerland), 13*(12). https://doi.org/10.3390/app13127167

Gulzar, M. W., Khan, M. K., Gulzar, R., Suleman, M., Hussain, J., Hassan, A., et al. (2024). Nanotechnology application in overcoming the reproductive disorders in livestock: Nanotechnology in livestock reproduction. *Letters In Animal Biology, 4*(2), 11–18.

Hammad, S. M., El-Wassefy, N. A., Shamaa, M. S., & Fathy, A. (2020). Evaluation of zinc-oxide nanocoating on the characteristics and antibacterial behavior of nickel-titanium alloy. *Dental Press Journal of Orthodontics, 25*(4), 51–58. https://doi.org/10.1590/2177-6709.25.4.051-058.oar

Hashem, N. M., & Gonzalez-Bulnes, A. (2020). State-of-the-art and prospective of nanotechnologies for smart reproductive management of farm animals. *Animals, 10*(5), 840.

Hashem, N. M., & Gonzalez-Bulnes, A. (2021). Nanotechnology and reproductive management of farm animals: Challenges and advances. *Animals, 11*(7), 1932.

Hill, E. K., & Li, J. (2017). Current and future prospects for nanotechnology in animal production. *Journal of Animal Science and Biotechnology, 8*, 1–13.

Hussain, S., Mubeen, I., Ullah, N., Shah, S. S. U. D., Khan, B. A., Zahoor, M., et al. (2022). Modern diagnostic imaging technique applications and risk factors in the medical field: A review. *BioMed Research International, 2022*. https://doi.org/10.1155/2022/5164970

Ianiski, L. B., de Rodrigues, F. S., et al. (2022). Nanotechnology in veterinary medicine: A review. *Ciência Rural, 52*(6). https://doi.org/10.1590/0103-8478cr20210195

Joudeh, N., & Linke, D. (2022). Nanoparticle classification, physicochemical properties, characterization, and applications: A comprehensive review for biologists. *Journal of Nanobiotechnology, 20*(1), 1–29. https://doi.org/10.1186/s12951-022-01477-8

Karuppannan, C., Sivaraj, M., Kumar, J. G., Seerangan, R., Balasubramanian, S., & Gopal, D. R. (2017). Fabrication of progesterone-loaded nanofibers for the drug delivery applications in bovine. *Nanoscale Research Letters, 12*, 1–6.

Khan, S., Ansari, Z. A., Alothman, O. Y., Fouad, H., & Ansari, S. G. (2017). Application of amine and copper doped magnesium oxide nanoparticles in electrochemical immunosensors for detecting Brucella abortus. *Nanoscience and Nanotechnology Letters, 9*(11), 1656–1664.

Khan, Y., Sadia, H., Zeeshan, S., Shah, A., Khan, M. N., Shah, A. A., et al. (2022). Nanoparticles, and their applications in various fields of nanotechnology: A review. *Catalysts, 12*(11), 1386.

Kuitio, C., Rasri, N., Kiriwan, D., Unajak, S., & Choowongkomon, K. (2020). Development of a biosensor from aptamers for detection of the porcine reproductive and respiratory syndrome virus. *Journal of Veterinary Science, 21*(5), e79.

Kurul, F., Turkmen, H., Cetin, A. E., & Topkaya, S. N. (2025). Nanomedicine: How nanomaterials are transforming drug delivery, bio-imaging, and diagnosis. *Next Nanotechnology, 7*(October 2024), 100129. https://doi.org/10.1016/j.nxnano.2024.100129

Labens, R., Daniel, C., Hall, S., Xia, X. R., & Schwarz, T. (2017). Effect of intra-articular administration of superparamagnetic iron oxide nanoparticles (SPIONs) for MRI assessment of the cartilage barrier in a large animal model. *PLoS One, 12*(12), 1–19. https://doi.org/10.1371/journal.pone.0190216

Li, K., Li, D., Li, C. H., Zhuang, P., Dai, C., Hu, X., et al. (2021). Efficient in vivo wound healing using noble metal nanoclusters. *Nanoscale, 13*(13), 6531–6537. https://doi.org/10.1039/d0nr07176e

Liang, A., Zhou, Y., Riaz, H., & Davis, J. S. (2023). Genetic analysis of reproductive traits in livestock. *Frontiers in Genetics, 13*, 1116038.

Linh, N. T., Qui, N. H., & Triatmojo, A. (2022). The effect of nano-encapsulated herbal essential oils on poultry's health. *Archives of Razi Institute, 77*(6), 2013.

Liu, F., Yao, Y., Zhu, B., Yu, Y., Ren, R., & Hu, Y. (2024). The novel imaging methods in diagnosis and assessment of cerebrovascular diseases: An overview. *Frontiers in Medicine, 11*(April), 1–14. https://doi.org/10.3389/fmed.2024.1269742

Lv, S., Sheng, J., Zhao, S., Liu, M., & Chen, L. (2018). The detection of brucellosis antibody in whole serum based on the low-fouling electrochemical immunosensor fabricated with magnetic Fe3O4@ Au@ PEG@ HA nanoparticles. *Biosensors and Bioelectronics, 117*, 138–144.

Ma, X., Tian, Y., Yang, R., Wang, H., Allahou, L. W., Chang, J., et al. (2024). Nanotechnology in healthcare, and its safety and environmental risks. *Journal of Nanobiotechnology, 22*(1). https://doi.org/10.1186/s12951-024-02901-x

Maki Baiee, R. S., Alaw Qotbi, A. A., & Hassan Habeeb, H. M. (2024). Effect of inorganic and Nano-zinc oxide on feed intake, weight gain and feed conversion of Awassi lambs. *Kufa Journal for Veterinary Medical Sciences, 15*(2).

Mancianti, F., & Ebani, V. V. (2020). Biological activity of essential oils. *Molecules, 25*(3), 2–5. https://doi.org/10.3390/molecules25030678

Mashweu, A. R., & Azov, V. A. (2024). Nanotechnology in drug delivery: Anatomy and molecular insight into the self-assembly of peptide-based hydrogels. *Molecules, 29*(23). https://doi.org/10.3390/molecules29235654

Mondal, R., Dam, P., Chakraborty, J., Paret, M. L., Katı, A., Altuntas, S., Sarkar, R., Ghorai, S., Gangopadhyay, D., Mandal, A. K., & Husen, A. (2022). Potencial del nanobiosensor en la agricultura sostenible: el estado del arte. *Heliyon, 8*(12), E12207. https://doi.org/10.1016/j.heliyon.2022.e12207

Montrose, A., Creedon, N., Sayers, R., Barry, S., & O'riordan, A. (2015). Novel single gold nanowire-based electrochemical Immunosensor for rapid detection of bovine viral Diarrhoea antibodies in serum. *J Biosens Bioelectron, 6*, 174. https://doi.org/10.4172/2155-6210.1000174

Moradi, M., Hajarian, H., Karamishabankareh, H., Soltani, L., & Soleymani, B. (2022). Pre-treatment of ram semen extender with magnetic nanoparticles on freeze-thawed spermatozoa. *Veterinary Medicine and Science, 8*(2), 792–798.

Naresh, V., & Lee, N. (2021). A review on biosensors and recent development of nanostructured materials-enabled biosensors. *Sensors (Basel)., 21*(4), 1109.

Neculai-Valeanu, A. S., Ariton, A. M., Mădescu, B. M., Rîmbu, C. M., & Creangă, Ş. (2021). Nanomaterials and essential oils as candidates for developing novel treatment options for bovine mastitis. *Animals, 11*(6), 1625.

Nehme, R., Andrés, S., Pereira, R. B., Ben Jemaa, M., Bouhallab, S., Ceciliani, F., et al. (2021). Essential oils in livestock: From health to food quality. *Antioxidants, 10*(2), 330.

Odhiambo, J. F., DeJarnette, J. M., Geary, T. W., Kennedy, C. E., Suarez, S. S., Sutovsky, M., & Sutovsky, P. (2014). Increased conception rates in beef cattle inseminated with nanopurified bull semen. *Biology of Reproduction, 91*(4), 97–91.

Oliveira, J. E., Medeiros, E. S., Cardozo, L., Voll, F., Madureira, E. H., Mattoso, L. H. C., & Assis, O. B. G. (2013). Development of poly (lactic acid) nanostructured membranes for the controlled delivery of progesterone to livestock animals. *Materials Science and Engineering: C, 33*(2), 844–849.

Paluszkiewicz, P., Martuszewski, A., Zaręba, N., Wala, K., Banasik, M., & Kepinska, M. (2022). The application of nanoparticles in diagnosis and treatment of kidney diseases. *International Journal of Molecular Sciences, 23*(1). https://doi.org/10.3390/ijms23010131

Patra, J. K., Das, G., Fraceto, L. F., Campos, E. V. R., Rodriguez-Torres, M. D. P., Acosta-Torres, L. S., et al. (2018). Nano based drug delivery systems: Recent developments and future prospects. *Journal of Nanobiotechnology, 16*(1), 1–33. https://doi.org/10.1186/s12951-018-0392-8

Paul, N., Kumaresan, A., Talluri, T. R., Raval, K., Elango, K., Pradeep Nag, B. S., et al. (2024). Lectin functionalised iron magnetic nanoparticle-based sperm selection: A potential technique to improve bull sperm quality in vitro. *Reproduction in Domestic Animals, 59*(10), e14733.

Poon, C., Gallo, J., Joo, J., Chang, T., Bañobre-López, M., & Chung, E. J. (2018). Hybrid, metal oxide-peptide amphiphile micelles for molecular magnetic resonance imaging of atherosclerosis. *Journal of Nanobiotechnology, 16*(1), 1–11. https://doi.org/10.1186/s12951-018-0420-8

Rabiei, P., Mohabatkar, H., & Behbahani, M. (2024). A label-free G-quadruplex aptamer/gold nanoparticle-based colorimetric biosensor for rapid detection of bovine viral diarrhea virus genotype 1. *PLoS One, 19*(7), e0293561.

Rios, T. B., Maximiano, M. R., Feitosa, G. C., Malmsten, M., & Franco, O. L. (2024). Nanosensors for animal infectious disease detection. *Sensing and Bio-Sensing Research, 43*, 100622.

Saliminasab, M., Jabbari, H., Farahmand, H., Asadi, M., Soleimani, M., & Fathi, A. (2022). Study of antibacterial performance of synthesized silver nanoparticles on Streptococcus mutans bacteria. *Nanomedicine Research Journal, 7*(4), 391–396. https://doi.org/10.22034/nmrj.2022.04.010

Shakweer, W. M. E., Azzaz, H. H., El-Nomeary, Y. A., El-Sayed, S. M., Youssef, A. M., & Hassaan, N. A. (2023). Synthesis and characterization of selenium nanoparticles and its effects on in vitro rumen feed degradation, ruminal parameters, and total gas production. *Egyptian Journal of Chemistry, 66*(12), 189–197.

Shao, K., Zhang, C., Ye, S., Cai, K., Wu, L., Wang, B., et al. (2017). Near–infrared electrochemiluminesence biosensor for high sensitive detection of porcine reproductive and respiratory syndrome virus based on cyclodextrin-grafted porous au/PtAu nanotube. *Sensors and Actuators B: Chemical, 240*, 586–594.

Sitaresmi, P. I., Hudaya, M. F., Atmoko, B. A., Wulandari, W., & Ujilestari, T. (2024). Production and effects of nanomineral selenium (Nano-Se) feed additive on rumen fermentation, productivity, and reproductive performance of ruminants. *Journal of Advanced Veterinary and Animal Research, 11*(3), 782.

Suminda, G. G. D., Bhandari, S., Won, Y., Goutam, U., Pulicherla, K. K., Son, Y. O., & Ghosh, M. (2022). High-throughput sequencing technologies in the detection of livestock pathogens, diagnosis, and zoonotic surveillance. *Computational and Structural Biotechnology Journal, 20*, 5378–5392.

Szczyglewska, P., Feliczak-Guzik, A., & Nowak, I. (2023). Nanotechnology–general aspects: A chemical reduction approach to the synthesis of nanoparticles. *Molecules, 28*(13). https://doi.org/10.3390/molecules28134932

Ta, H. T., Li, Z., Hagemeyer, C. E., Cowin, G., Zhang, S., Palasubramaniam, J., et al. (2017). Molecular imaging of activated platelets via antibody-targeted ultra-small iron oxide nanoparticles displaying unique dual MRI contrast. *Biomaterials, 134*, 31–42. https://doi.org/10.1016/j.biomaterials.2017.04.037

Tu, Y., Ma, X., Chen, H., Fan, Y., Jiang, L., Zhang, R., & Cheng, Z. (2022). Molecular imaging of matrix Metalloproteinase-2 in atherosclerosis using a smart multifunctional PET/MRI nanoparticle. *International Journal of Nanomedicine, 17*(November), 6773–6789. https://doi.org/10.2147/ijn.s385679

Wahab, R., Khan, S. T., Ahmad, J., Musarrat, J., & Al-Khedhairy, A. A. (2017). Functionalization of anti-Brucella antibody on ZnO-NPs and their deposition on aluminum sheet towards developing a sensor for the detection of Brucella. *Vacuum, 146*, 592–598.

Williams, A., Bennison, J. J., Mackenzie, A. M., & Sinclair, L. A. (2024). Feeding nanoparticles of copper oxide coated with lysine with or without added antagonists affects the copper status but not the performance of Holstein dairy cows. *Journal of Dairy Science, 107*(11), 9277–9288.

Yagublu, V., Karimova, A., Hajibabazadeh, J., Reissfelder, C., Muradov, M., Bellucci, S., & Allahverdiyev, A. (2022). Overview of physicochemical properties of nanoparticles as drug carriers for targeted cancer therapy. *Journal of Functional Biomaterials, 13*(4). https://doi.org/10.3390/jfb13040196

Yang, B., Zeng, X., Ge, Y., Liu, W., Hao, W., Long, C., et al. (2024). A new method for rapid, portable, low-cost detection of Clostridium perfringens β2 toxin in animal fecal using smartphone-based electrochemical immunosensor. *Microchemical Journal, 198*, 110138.

Yen, C. C., Chen, Y. C., Wu, M. T., Wang, C. C., & Wu, Y. T. (2018). Nanoemulsion as a strategy for improving the oral bioavailability and anti-inflammatory activity of andrographolide. *International Journal of Nanomedicine, 13*, 669–680. https://doi.org/10.2147/IJN.S154824

Yousef, M. S., López-Lorente, A. I., Diaz-Jimenez, M., Consuegra, C., Dorado, J., Pereira, B., et al. (2020). Nano-depletion of acrosome-damaged donkey sperm by using lectin peanut agglutinin (PNA)-magnetic nanoparticles. *Theriogenology, 151*, 103–111.

Yousefdoost, S., Samadi, F., Jafari, S. M., Ramezanpour, S. S., Hassani, S., & Ganji, F. (2019). Application of nanoencapsulated silymarin to improve its antioxidant and hepatoprotective activities against carbon tetrachloride-induced oxidative stress in broiler chickens. *Livestock Science, 228*, 177–186.

Yousefi, M., Adineh, H., Hoseini, S. M., Hashemianfar, S. A. M., Kulikov, E. V., Petukhov, N. V., & Ryzhova, T. A. (2025). Effects of Laurus nobilis essential oil nano-particles on growth performance, antioxidant and immune responses to bacterial infection in Nile tilapia, Oreochromis niloticus. *Aquaculture, 596*, 741821.

Yu, M., Fu, S., Li, J., Yi, X., Qu, Z., Li, Y., et al. (2023). A smartphone-based portable biosensor for colorimetric detection of porcine reproductive and respiratory syndrome virus. *Sensors and Actuators B: Chemical, 396*, 134617.

Zeng, J., Wu, J., Li, M., & Wang, P. (2018). A novel magnetic nanoparticle for early detection of amyloid plaques in Alzheimer's disease. *Archives of Medical Research, 49*(4), 282–285. https://doi.org/10.1016/j.arcmed.2018.09.005

Zhang, S., Zhou, H., Kong, N., Wang, Z., Fu, H., Zhang, Y., et al. (2021). L-cysteine-modified chiral gold nanoparticles promote periodontal tissue regeneration. *Bioactive Materials, 6*(10), 3288–3299. https://doi.org/10.1016/j.bioactmat.2021.02.035

Zhao, Y., Zhang, L., Wang, A., & Zhou, D. (2025). Tecnología de biosensores: avances y aplicaciones en el diagnóstico de enfermedades infecciosas en el ganado. *Ciencias Veterinarias, 12*(1), 23. https://doi.org/10.3390/vetsci12010023

Zhu, Q., Chai, Y., Zhuo, Y., & Yuan, R. (2015). An amperometric immunosensor for detection of Streptococcus suis serotype 2 using a nickel–gold nanocomposite as a tracer matrix. *RSC Advances, 5*(97), 79323–79328.

Zhu, X., Xiong, H., Zhou, Q., Zhao, Z., Zhang, Y., Li, Y., et al. (2021). A pH-Activatable MnCO3Nanoparticle for improved magnetic resonance imaging of tumor malignancy and metastasis. *ACS Applied Materials and Interfaces, 13*(16), 18462–18471. https://doi.org/10.1021/acsami.0c22624

Zou, X., Huang, H., Gao, Y., & Su, X. (2012). Detection of avian influenza virus based on magnetic silica nanoparticles resonance light scattering system. *Analyst, 137*(3), 648–653.

Chapter 9
Phytotherapeutic Alternatives: New Nanotechnology-Based Approaches to Tick Control

Romario García-Ponce, Adriana E. Flores-Suarez, Jesús Jaime Hernández-Escareño, Gustavo Hernández Vidal, José Pablo Villarreal Villarreal, and Luis Humberto Díaz García

1 Introduction

Ticks are a global public and veterinary health problem, affecting both human populations and domestic animals. These blood-sucking ectoparasites constitute the most diverse group of disease-carrying arthropods, ranking second only to mosquitoes in importance as transmitters of pathogens of medical and veterinary importance (Benelli et al., 2016).

Currently, a significant increase in the global prevalence of ticks and the diseases they transmit has been documented, both in humans and in animals. This phenomenon appears to be associated with multiple factors, including changes in human behavior patterns, demographic dynamics, climatic variations, and land-use changes. These conditions have created an environment conducive to the emergence and re-emergence of vector-borne infectious and zoonotic diseases (Charles et al., 2021). Among the pathogens they transmit are the causative agents of anaplasmosis, babesiosis, ehrlichiosis, rickettsiosis, borreliosis, as well as various viral diseases (Muqaddas et al., 2025). In addition to their role as vectors, ticks also have a considerable economic impact on livestock-production systems. In particular, the species *Rhipicephalus microplus* is estimated to be responsible for economic losses of US$ 573.61 million in Mexico (Rodríguez-Vivas et al., 2017) and US$ 3.24 billion

R. García-Ponce · A. E. Flores-Suarez
Faculty of Biological Sciences, Autonomous University of Nuevo León, San Nicolás de los Garza, Mexico

J. J. Hernández-Escareño · G. H. Vidal · J. P. V. Villarreal (✉)
Faculty of Veterinary Medicine and Zootechnics, Autonomous University of Nuevo León, General Escobedo, Mexico

L. H. D. García
Unidad Académica de Medicina Veterinadia y Zootecnia, Universidad Autónoma de Zacatecas, Zacatecas, Mexico

M. S. Heya (ed.), *Natural Alternatives and Nanotechnology Applied to the One Health Approach*, https://doi.org/10.1007/978-3-032-03675-9_9

in Brazil (Grisi et al., 2014). These data demonstrate the economic impact of ticks in countries with high livestock production.

In recent decades, conventional control of tick infestations has predominantly relied on the use of synthetic ixodicides. However, their continuous, indiscriminate application, often lacking adequate technical guidance, has significantly contributed to the development of tick populations resistant to commonly used active ingredients (De Meneghi et al., 2016; Rodriguez-Vivas et al., 2021; Valsoni et al., 2020, 2021; Vilela et al., 2020). This situation is further exacerbated by the detection of ixodicide residues in animal products, such as meat and milk, as well as by the adverse effects of these compounds on the environment, which is a concern from the perspective of public health and environmental sustainability (Higa et al., 2016; Jain et al., 2020; Rodriguez-Vivas et al., 2011). In this context, research and development of effective alternatives for tick control are necessary, ensuring a path toward more sustainable and responsible solutions in pest control (García-Ponce et al., 2025).

Plants represent an important source of bioactive compounds with diverse biological and pharmacological properties, which positions them as important elements in the search and development of new therapeutic agents (Tienda-Vázquez et al., 2023). These secondary metabolites of plant origin also stand out for their high safety profile for both human health and the environment (Pavela, 2014). Currently, the research of phytotherapeutic alternatives with insecticidal and acaricidal activity constitutes an emerging trend in tick control, showing promising results (Adenubi et al., 2016; Molaba et al., 2025). Plant compounds can act on multiple organs and systems in ticks (García-Ponce et al., 2025), suggesting the presence of diverse mechanisms of action that not only increase their efficacy, but could also help to mitigate the risk of resistance development (Chagas, 2015; Pavela & Benelli, 2016b). However, the practical application of these compounds under field conditions presents certain challenges, among which the need to formulate suitable vehicles that allow their solubilization, stabilization, and protection against degradation processes induced by environmental factors, such as photooxidation, pH variations, thermal fluctuations, and microbial activity, stands out (Benelli et al., 2017; Mulla & Su, 1999; Pavela & Benelli, 2016a).

In this context, nanotechnology can provide solutions to the limitations associated with the use of plant compounds, through the development of nanometric systems capable of incorporating active ingredients from plants (Figueiredo et al., 2022; García-Ponce et al., 2025). Nanoparticles (NPs) have proven to be effective vehicles for the encapsulation of natural products, offering multiple advantages, such as greater physicochemical stability, protection against environmental degradation processes, the ability to cross various biological barriers, and a controlled and sustained release of active ingredients (Armendáriz-Barragán et al., 2016; Berni et al., 2022; Carvalho et al., 2020; Figueiredo et al., 2022; Irache et al., 2011). The innovative use of NPs to deliver bioactive compounds also eliminates the need to use toxic solubilizing agents, allowing their ixodicidal activity to be exploited more safely (Carvalho et al., 2020; Ventola, 2017).

In this chapter, we focus on describing the effects of naturally derived products incorporated into NPs that have been evaluated against different tick species, discussing several challenges and prospects for this novel field of research.

2 Nanotechnology in Tick Control

Nanotechnology is a discipline that studies, designs, and applies materials and systems at the nanoscale. Their unique physicochemical properties allow for the optimization of various physical, chemical, and biological processes (Riehemann et al., 2009). In the case of controlled-release systems, these properties are closely linked to their small size, which confers a high surface-to-volume ratio that favors their interaction with biological structures and improves the bioavailability of active ingredients (Zainab et al., 2024).

Among the NPs explored for the delivery of natural compounds with ixodicidal activity, polymeric NPs, lipid NPs, and nanoemulsions stand out (Fantatto et al., 2024; Figueiredo et al., 2022, 2023; García-Ponce et al., 2025; Ibrahium et al., 2024; Nogueira et al., 2020; Pazinato et al., 2014). The general characteristics of these NPs are illustrated in Fig. 9.1 (Duan et al., 2020; Gupta et al., 2019; Kreuter, 2007; Lammari et al., 2020; Pardeike et al., 2009; Salvia-Trujillo et al., 2017).

Nanometric systems allow for the efficient encapsulation of active ingredients with both hydrophilic and hydrophobic ixodicidal activity, optimizing their physicochemical stability and bioavailability. Their development enables controlled and targeted release of the compound, offering protection against adverse environmental factors, promoting intracellular penetration and distribution, and promoting

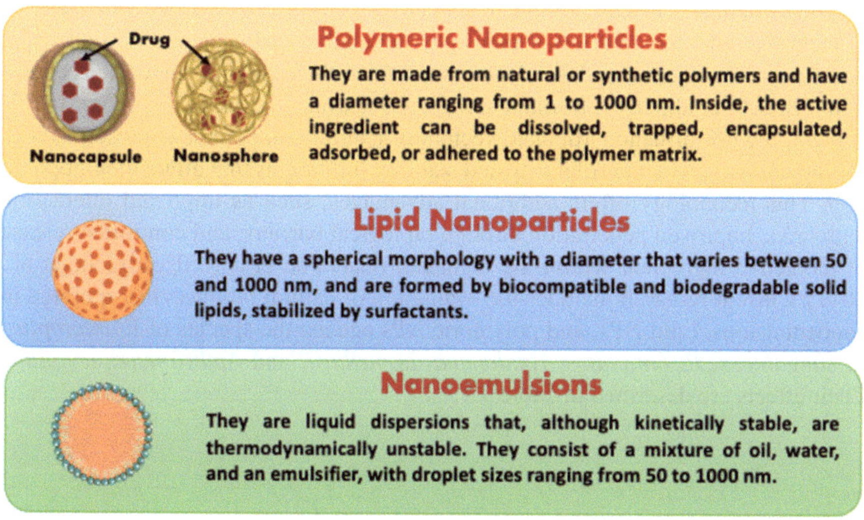

Fig. 9.1 Nanoparticle systems used in controlling ticks

selective accumulation of the active ingredient in target tissues or cells (Carvalho et al., 2020; Couvreur & Vauthier, 2006; Hillaireau & Couvreur, 2009; Irache et al., 2011). Furthermore, they present unique characteristics that are beneficial in the development of tick treatments, since they contribute to reducing dose-dependent toxicity and eliminating the need to use toxic solvents that are usually required for their conventional formulation (Carvalho et al., 2020; Ventola, 2017). The versatility in the structural design of these systems, which includes control of size, morphology, surface charge, and release profile, allows a specific and efficient interaction with tissues, cells, or molecular structures of the target organism, increasing their therapeutic efficacy (Shahzad & Manzoor, 2019).

3 Phytotherapeutic Alternatives Based on Nanoparticles for Tick Control

The use of phytotherapeutic alternatives is a growing trend in the development of treatments for tick control (Molaba et al., 2025). Exploring the ixodicidal activity of extracts and essential oils has allowed the identification of naturally occurring molecules with potential effects for use as ixodicidal treatments. This is because they are considered safer than synthetic ixodicides and are valued for their potential to reduce chemical residues in food and minimize risks to human health and the environment (Pavela, 2014). Furthermore, plant compounds, products of secondary plant metabolism, have the ability to interfere with tick organs and systems (Fig. 9.2), exerting multiple mechanisms of action. This functional diversity not only contributes to efficacy against the ectoparasite but may also reduce the risk of resistance development (Chagas, 2015; Pavela and Benelli, 2016b). However, their practical application faces significant challenges, particularly related to the stability of natural compounds. Factors such as photooxidation, pH changes, thermal variations, and microbial degradation can compromise their efficacy under field conditions (Benelli et al., 2017; Mulla and Su, 1999; Pavela and Benelli, 2016a). In this context, the use of nanoparticulate vectors of natural compounds is presented as a strategy to overcome the solubility and stability limitations that affect their application. This technology offers additional advantages, such as improved interaction with ticks, improved penetration through biological barriers, and controlled release of the active ingredient within the organism (Carvalho et al., 2020; Irache et al., 2011). Current research includes the evaluation of natural products incorporated in nanoemulsions, lipid NPs, and polymeric NPs against the species of *R. microplus*, *R. sanguineus*, *R. annulatus*, *Hyalomma dromedarii*, and *Amblyomma sculptum*. Their effects are described in Table 9.1.

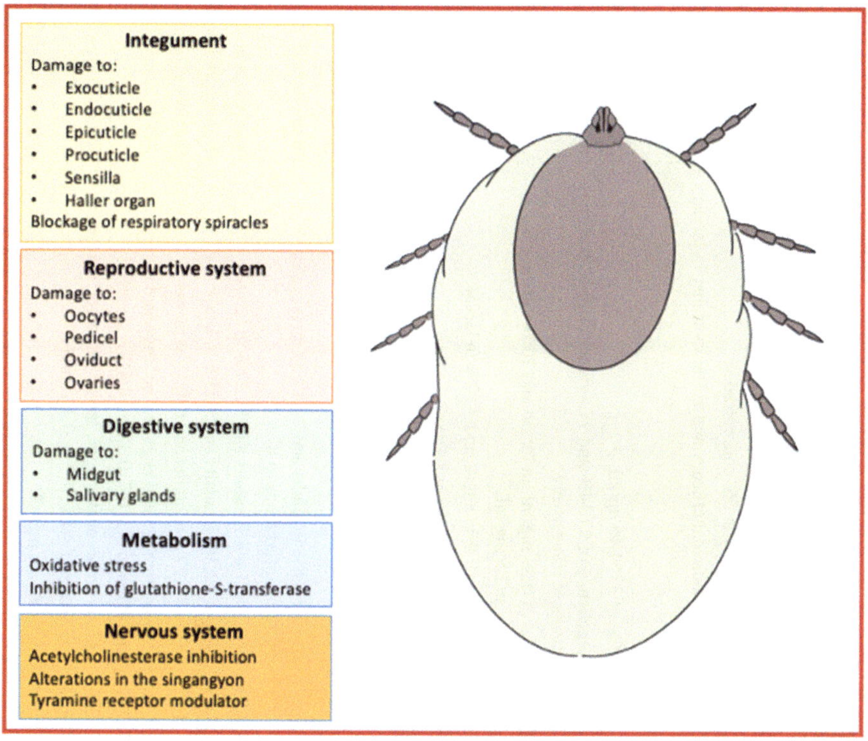

Fig. 9.2 Biological targets and impacts of natural compounds on ticks

3.1 Nanoemulsions

Recent research has highlighted the efficacy of using essential oils incorporated into nanoemulsions, particularly those derived from species such as *Melaleuca alternifolia*, *Eucalyptus globulus*, *Cinnamomum verum*, *C. zeylanicum*, *C. camphora*, *Cuminum cyminum*, *Pimenta dioica*, *Pilocarpus spicatus*, *Pelargonium graveolens*, *Origanum majorana*, *Allium sativum*, *Citrus* × *limon*, *Artemisia herba-alba*, and *Melia azedarach*. Furthermore, active compounds present in these oils, such as d-limonene, thymol and 6,7,8,9-tetrahydrospilanthol, have been evaluated for their ixodicidal and repellent properties against different tick species such as *Rhipicephalus microplus*, *R. sanguineus*, *R. annulatus*, *Hyalomma dromedarii*, and *Amblyomma sculptum*. Studies report mortality rates ranging from 60% to 100%, as well as repellency levels of up to 100%. Ixodicidal activity has been determined by calculating the median lethal concentration (LC_{50}), the values of which are summarized in Table 9.1. In some cases, the incorporation of essential oils and volatile compounds into nanoemulsions has significantly improved their biological efficacy against ticks. However, some studies have reported a decrease in ixodicidal activity

Table 9.1 Nanoparticles evaluated against ticks: composition, characteristics, and effects

Nanoparticles	Tick species	Composition	Size (nm)	Test	Summary of results	References
Nanoemulsions	*Rhipicephalus microplus*	*Cinnamomum verum* EO	175.6	AIT, in vivo	Ixodicidal efficacy of 97% and protection from natural infestations	dos Santos et al. (2017)
	R. microplus	*Eucalyptus globulus* EO	69 ± 0.71	AIT	61.2% efficacy at 5% concentration	Galli et al. (2018)
	R. microplus	*C. zeylanicum*, *Cuminum cyminum*, and *Pimenta dioica* EO	856.8 ± 216.4	AIT, LPT	100% ixodicidal efficacy and 100% mortality in larvae	Lazcano Díaz et al. (2019)
	R. microplus	*Cinnamomum camphora* EO	101.6	LPT	LC_{50} of 3364.46 μg/mL	Kapoor and Preet (2023)
	R. microplus	*Pilocarpus spicatus* EO	86.06 ± 0.2916	RT	Repellency of 97.14% and 97.89% at 6 and 10 h after treatment respectively	Nogueira et al. (2020)
	R. microplus	6,7,8,9-Tetrahydrospilanthol	145–175	LPT, AIT	LC_{50} of 0.39 mg/mL in larvae and 100% efficacy in adults at 2.5 mg/mL	Anholeto et al. (2024)
	R. sanguineus	d-Limonene	258.5	AIT, LPT	LC_{50} of 3.97% in adults and 7.82% in larvae	Ibrahium et al. (2024)
	R. sanguineus	*Pelargonium graveolens* EO + *Origanum majorana* EO + thymol	Unknown	In vivo	100% effectiveness at 14 days	Gadelhaq et al. (2024)
	R. annulatus	*Allium sativum EO*	32.2–96.3	AIT	100% mortality at 3 mg/L	Abd-Elrahman et al. (2024)
	R. annulatus	*Pelargonium graveolens EO*	17.84	AIT, LPT, in vivo	LC_{50} of 5.6% in adults and 1.68% in larvae, also reducing infestations to 87.97%	Ibrahium et al. (2022a)
	R. annulatus	*Citrus x limon EO*	50.66	AIT, LPT	LC_{50} of 11.4% in adults and 32.27% in larvae	Ibrahium et al. (2022b)
	R. annulatus	*Melaleuca alternifolia* oil	188.4	AIT, LPT	LC_{50} of 12.3% in adults and 3.55% in larvae	Ibrahium et al. (2022b)
	R. annulatus	d-Limonene	258.5	AIT, LPT	LC_{50} of 1.27% in adults and 6.48% in larvae	Ibrahium et al. (2024)
	Hyalomma dromedarii	*Artemisia herba-alba* extract	62–69	EIT, LIT, NIT, AIT	LC_{50} in embryonated eggs, larvae, engorged nymphs, and adults of 0.3, 0.7, 0.3, and 4.4% respectively	Abdel-Ghany et al. (2021)
	H. dromedarii	*Melia azedarach* extract	52–91	EIT, LIT, NIT, AIT	LC_{50} in embryonated eggs, larvae, engorged nymphs, and adults of 1.1%, 1.7%, 0.4%, and 22.2% respectively	Abdel-Ghany et al. (2021)
	Amblyomma sculptum	6,7,8,9-Tetrahydrospilanthol	145–175	LIT	LC_{50} of 1.36 mg/mL	Anholeto et al. (2024)

Lipid nanoparticles	*R. microplus*	*M. alternifolia* essential oils	287 ± 2	AIT	100% reproductive inhibition	Pazinato et al. (2014)
	R. microplus	*M. alternifolia* essential oils	287 ± 2	AIT, in vivo	Reproductive inhibition of 34.5% and protection for 4 days post-treatment	Pazinatto Boito et al. (2016)
	R. microplus	Cypermethrin + chlorpyrifos + citral	286–304	LPT	LC_{50} of 3.3 μg/mL	Figueiredo et al. (2022)
	R. microplus	Cypermethrin + chlorpyrifos + menthol	286–304	LPT	LC_{50} of 5.4 and 2.3 μg/mL for SLN and NLC respectively	
	R. microplus	Cypermethrin + chlorpyrifos + limonene	286–304	LPT	LC_{50} of 4 and 5.5 μg/mL for SLN and NLC respectively	
	R. annulatus	Carvacrol + cineol + cholesterol + phospholipid	267 ± 2.25	AIT, LPT	In adults, the LC_{50} was 2.6% and the LC_{90} was 3.84%; in larvae, the LC_{50} was 0.23% and the LC_{90} was 0.49%	Gamal et al. (2023)
	R. sanguineus	Carvacrol + cineol + cholesterol + phospholipid	267 ± 2.25	LPT	CL_{50} of 0.21% and CL_{90} of 0.48%	Gamal et al. (2023)
Polymeric nanoparticles	*R. microplus*	*Pouteria gardneriana* extract + chitosan	589	AIT	Reduction of egg mass and mortality of the extract to 0.4% with 0.5 g of chitosan nanoparticles	Barbosa et al. (2016)
	R. microplus	*C. verum* essential oil + poly-Ɛ-caprolactone	189.5	AIT, in vivo	95% ixodicidal efficacy and protection from natural infestations	dos Santos et al. (2017)
	R. microplus	*Eucalyptus globulus* EO	62.1 ± 1.25	AIT	35.5% efficacy at 5% concentration	Galli et al. (2018)
	R. microplus	Cypermethrin + chlorpyrifos + citral + zeina	290.64	LPT, AIT	Mortality in larvae of 100% and ixodicidal efficacy of 50.2% in adults	Figueiredo et al. (2023)
	R. microplus	Cypermethrin + chlorpyrifos + menthol + zeina	287.12	LPT, AIT	Mortality in larvae of 100% and ixodicidal efficacy of 40.5% in adults	
	R. microplus	Cypermethrin + chlorpyrifos + limonene + zeina	282.18	LPT, AIT	Mortality in larvae of 100% and ixodicidal efficacy of 60.1% in adults	
	R. microplus	*Achyrocline satureioides* extract + Poly-Ɛ-caprolactone	192.33 ± 4.35	LPT	Mortality of 4.38 ± 7.1% at 20 mg/mL	Fantatto et al. (2024)
	R. microplus	*Eucalyptus globulus* EO + chitosan	25–30	RT	Repellency of up to 90% was observed	Nawaz et al. (2025)
	R. sanguineus	Thymol + Poly-Ɛ-caprolactone	195.7 ± 0.5	RT	60–70% repellency for 2 h against nymphs	Sales et al. (2024)

LPT larval pack test, *AIT* adult immersion test, *EIT* egg immersion test, *NIT* nymph dive test, *RT* repellence test, *EO* essential oil, LC_{50} lethal concentration 50%, *SLN* solid lipid nanoparticles, *NLC* nanostructured lipid carriers compared with free compounds, which could be attributed to a delayed release of the active ingredient from the nanoemulsions (dos Santos et al., 2017).

The effectiveness of these oils is associated with their diverse range of bioactive compounds, capable of acting through multiple mechanisms on different biological targets (Lazcano Díaz et al., 2019). Furthermore, their formulation in nanoemulsions improves the stability, bioavailability, and persistence of the compounds, increasing their effectiveness. These characteristics position nanoemulsions as promising tools within integrated control strategies, offering a sustainable and efficient alternative for tick management.

3.2 Lipid Nanoparticles

Lipid NPs have also been shown to be effective against *R. microplus*, *R. annulatus*, and *R. sanguineus* species, in research conducted by Pazinato et al. (2014) and Pazinatto Boito et al. (2016), highlighting the incorporation of *M. alternifolia* essential oil, which increased ixodicidal activity, achieving 100% reproductive inhibition in female ticks. In both investigations, the effects of *M. alternifolia* essential oil were significantly enhanced when incorporated into these nanocarriers. Following the same line of thought, solid lipid NPs and nanostructured lipid carriers incorporating a combination of synthetic ixodicides (cypermethrin, chlorpyrifos) and natural products (terpenes; citral, menthol, and limonene) have been evaluated, showing LC_{50} between 3.3 and 5.5 μg/mL in *R. microplus* larvae. These results demonstrated the ability of lipid NPs to incorporate synthetic ixodicides and natural products. Furthermore, the effects of these formulations were greater than those containing only cypermethrin and chlorpyrifos (Figueiredo et al., 2022). In the case of *R. annulatus* and *R. sanguineus* species, lipid NPs loaded with carvacrol, and cineole have been evaluated, with mortality rates of up to 100%. These formulations also inhibited acetylcholinesterase activity and decreased glutathione and malonedealdehyde levels in treated ticks. The conclusion is that the nanoformulations significantly improve the adulticidal and repellent activities of carvacrol against both tick species (Gamal et al., 2023).

Lipid NPs as vehicles for the delivery of naturally derived ixodicidal products offer several advantages. Owing to their nanometric size and lipophilic composition, these NPs more easily cross biological barriers, which favors the absorption of active compounds and enhances their ixodicidal activity. Furthermore, its high adhesion and retention capacity on the cuticular surface of arthropods contributes to prolonging the contact time, thus increasing the effectiveness of the treatment against ticks (Tortorici et al., 2022).

3.3 Polymeric Nanoparticles

Polymeric NPs loaded with natural products have been explored as ixodicides and repellents in the fight against ticks. In particular, dos Santos et al. (2017) incorporated cinnamon oil into poly-Ɛ-caprolactone NPs, achieving 95% efficacy on adult female *R. microplus* ticks. These results were significantly higher than those shown by the essential oil in its pure form. Similarly, Figueiredo et al. (2023) showed that zein NPs formulated with combinations of synthetic ixodicides (cypermethrin, chlorpyrifos) and terpenes (citral, menthol, and limonene) showed mortality rates greater than 80% at concentrations above 0.029 mg/mL. Furthermore, efficacy rates of between 40.5% and 60.1% were obtained at concentrations of 0.466 mg/mL in engorged adult females. In both tests, the ixodicidal activity of the nanoformulations was greater than that of the commercial product Colosso®, which contains cypermethrin, chlorpyrifos, and citronellal. Furthermore, these formulations remained stable during storage and were shown to reduce the toxicity of the incorporated compounds in *Caenorhabditis elegans*, a nontarget organism. Repellent activity was reported by Sales et al. (2024), who developed formulations of thymol-loaded poly-β-caprolactone NPs and obtained 60–70% repellency for 2 h against *R. sanguineus* nymphs. Similarly, Nawaz et al. (2025) developed chitosan NPs loaded with *Eucalyptus globulus* essential oil, showing repellency of up to 90% against *R. microplus* larvae.

Polymeric NPs have the advantage of being formulated with biodegradable and biocompatible polymers, which allows their physicochemical characteristics to be adjusted to optimize the release of active ingredients based on specific environmental and biological conditions (García-Ponce et al., 2025). This versatility translates into greater availability and absorption of the active compounds, increasing their therapeutic efficacy. Furthermore, their formulation in aqueous systems favors the application of natural extracts, facilitating their integration into integrated pest management programs (Elmowafy et al., 2023). The encapsulation of natural products in polymeric matrices constitutes a promising strategy for the development of more stable, sustainable, and environmentally friendly treatments, with the potential to reduce the emergence of resistance to synthetic ixodicides. However, further progress is needed both in the optimization of these nanoformulations and in the design of standardized methodologies to validate their effectiveness against ticks under field conditions.

4 Nanoparticles/Tick Interactions

Effective tick treatments depend largely on the application method. Conventional techniques such as manual spraying, immersion baths, topical application, and parenteral administration determine the route of interaction of the active compound with the ectoparasite. NPs, being generally suspended in aqueous solutions, are compatible with these techniques without the need for adaptations. Their main routes of interaction with ticks include direct contact, ingestion, and inhalation (Shahzad & Manzoor, 2019).

The tegument, composed of an epicuticle and a procuticle (porous exocuticle and lamellar endocuticle), acts as the main barrier against topical agents (Kaufman, 2013; Remedio et al., 2015). Thanks to their nanometric size, NPs can pass through the pores of the exocuticle, which measure between 100 and 130 nm in some species (Axtell et al., 1971). The digestive and respiratory tracts also allow the entry of NPs, provided that they are present in the food or come into contact with the respiratory spiracles (García-Ponce et al., 2025).

Once they overcome external barriers, NPs interact efficiently with cellular structures. Their small size and large surface area favor tissue accumulation and cellular internalization, generally through endocytosis (Sun et al., 2019). This process, which may involve receptor-specific interactions or electrostatic attraction, leads to the formation of vesicles that fuse with lysosomes, facilitating the intracellular release of the active ingredient.

The mechanism of endocytosis varies depending on the properties of the NPs (size, shape, charge, functionalization) and the cell type. It can include pinocytosis or phagocytosis, the latter commonly associated with hemocytes, although not exclusively (Heya et al., 2023; Shahzad & Manzoor, 2019). Phagocytosis allows the internalization of larger particles, contributing to the intracellular distribution of NPs (Alves et al., 2021; Hillaireau & Couvreur, 2009).

5 Conclusions and Perspectives

Nanotechnology is revolutionizing the design of tick treatments by enabling more targeted delivery and release of naturally occurring active ingredients. These formulations offer advantages over conventional ixodicides, including overcoming resistance, reducing chemical residues in food, and lowering environmental impact. The physicochemical properties of NPs, as well as their ability to cross biological barriers and release active ingredients, have been highlighted as key factors in improving therapeutic efficacy and reducing side effects. Furthermore, the combination of phytotherapeutic alternatives and synthetic ixodicides in a single formulation allows for synergies that strengthen treatment efficacy and reduce the risk of resistance (Fantatto et al., 2025; Farina et al., 2024; García-Ponce et al., 2025).

Despite their potential, there are currently no nanoixodicide products on the market owing to technical, regulatory, and economic challenges. These include the need to ensure the safety and biocompatibility of NPs, reduce production costs, and scale manufacturing processes to industrial levels (Farjadian et al., 2019; Peltonen, 2018). The transition from laboratory to industrial production requires time, investment, and compliance with high-quality standards.

The registration of these products faces additional obstacles, given that nanoformulations must comply with regulations designed for conventional compounds, without standardized analytical protocols to assess their quality, safety, and efficacy (Carvalho et al., 2020; Ventola, 2017). The FDA has highlighted the need to develop appropriate regulatory frameworks and warned about the limited evidence available regarding the effects of the physicochemical properties of NPs on their safety (Paradise, 2019; Tinkle et al., 2014).

In conclusion, although nanotechnology represents a promising avenue for the development of innovative ixodicides of natural origin, its commercialization involves overcoming important scientific, regulatory, and economic challenges, which can prolong the process for up to 10–14 years (Carvalho et al., 2020; Zolnik et al., 2011).

References

Abdel-Ghany, H. S. M., Abdel-Shafy, S., Abuowarda, M., El-Khateeb, R. M., Hoballah, E. M., & Fahmy, M. M. (2021). Acaricidal activity of *Artemisia herba-alba* and *Melia azedarach* oil nanoemulsion against *Hyalomma dromedarii* and their toxicity on Swiss albino mice. *Experimental and Applied Acarology, 84*(1), 241–262. https://doi.org/10.1007/s10493-021-00618-2

Abd-Elrahman, S. M., Dyab, A. K., Kamel, F. A., Khedr, A. A., Khalifa, M. M., Mohamed, S. M., & Abdel-Hakeem, S. S. (2024). Assessment of cattle tick infestation: Molecular insights into *Rhipicephalus annulatus* and the efficacy of garlic oil and nanoemulsion as acaricidal agents. *Veterinary Parasitology, 329*, 110211. https://doi.org/10.1016/J.VETPAR.2024.110211

Adenubi, O. T., Fasina, F. O., McGaw, L. J., Eloff, J. N., & Naidoo, V. (2016). Plant extracts to control ticks of veterinary and medical importance: A review. *South African Journal of Botany, 105*, 178–193. https://doi.org/10.1016/J.SAJB.2016.03.010

Alves, M. S. R., Reis, A. A. L., da Silva Rocha, M. B., Alves, M. C. C., Borges, D. A., de Avelar, B. R., Campos, D. R., Cid, Y. P., & Scott, F. B. (2021). Cellular and metabolic response of hemolymph from engorged *Rhipicephalus microplus* females exposed to fluazuron. *Revista Brasileira de Medicina Veterinaria, 43*. https://doi.org/10.29374/2527-2179.BJVM001821

Anholeto, L. A., Brancaglion, G. A., et al. (2024). Acaricidal activity of synthetic spilanthol derivative against ticks of medical and veterinary importance. *Veterinary Parasitology, 327*, 110137. https://doi.org/10.1016/J.VETPAR.2024.110137

Armendáriz-Barragán, B., Zafar, N., Badri, W., Galindo-Rodríguez, S. A., Kabbaj, D., Fessi, H., & Elaissari, A. (2016). Plant extracts: From encapsulation to application. In *Expert opinion on drug delivery* (Vol. 13, pp. 1165–1175). Taylor and Francis Ltd.. https://doi.org/10.1080/17425247.2016.1182487

Axtell, R. C., Foelix, R. F., Coons, L. B., & Roshdy, M. A. (1971). Sensory receptors in ticks and mites. In *Proceeding of the 3rd international congress of acarology* (pp. 35–36).

Barbosa, P. F. P., et al. (2016). Application of polymeric nanoparticles for controlled release of ethanolic extract of guapeva leaves (*Pouteria gardneriana* Radlk) against *Riphicephalus (Boophilus) microplus* through in vitro studies. *African Journal of Biotechnology, 15*(49), 2778–2786. https://doi.org/10.5897/AJB2016.15230

Benelli, G., Pavela, R., Canale, A., & Mehlhorn, H. (2016). Tick repellents and acaricides of botanical origin: A green roadmap to control tick-borne diseases? *Parasitology Research, 115*(7), 2545–2560. https://doi.org/10.1007/S00436-016-5095-1

Benelli, G., Pavela, R., Maggi, F., Petrelli, R., & Nicoletti, M. (2017). Commentary: Making green pesticides greener? The potential of plant products for Nanosynthesis and Pest control. *Journal of Cluster Science, 28*(1), 3–10. https://doi.org/10.1007/s10876-016-1131-7

Berni, E., de Barbosa, R. M., & Durán, N. (2022). Chitosan-coated poly (Ɛ-caprolactone) nanoparticles as acaricide carriers. *Ticks and Tick-borne Diseases, 13*(1). https://doi.org/10.1016/j.ttbdis.2021.101849

Carvalho, S. G., Araujo, V. H. S., dos Santos, A. M., Duarte, J. L., Silvestre, A. L. P., Fonseca-Santos, B., Villanova, J. C. O., Gremião, M. P. D., & Chorilli, M. (2020). Advances and challenges in nanocarriers and nanomedicines for veterinary application. *International Journal of Pharmaceutics, 580*. https://doi.org/10.1016/j.ijpharm.2020.119214

Chagas, A. C. S. (2015). Medicinal plant extracts and nematode control. *CABI Reviews, 2015*, 1–8. https://doi.org/10.1079/PAVSNNR201510008

Charles, R. A., Bermúdez, S., Banović, P., Alvarez, D. O., Díaz-Sánchez, A. A., Corona-González, B., Etter, E. M. C., González, I. R., Ghafar, A., Jabbar, A., Moutailler, S., & Cabezas-Cruz, A. (2021). Ticks and tick-borne diseases in Central America and the Caribbean: A one health perspective. *Pathogens, 10*(10), 1273. https://doi.org/10.3390/PATHOGENS10101273

Couvreur, P., & Vauthier, C. (2006). Nanotechnology: Intelligent design to treat complex disease. *Pharmaceutical Research, 23*(7), 1417–1450. https://doi.org/10.1007/s11095-006-0284-8

de Higa, L. O. S., Garcia, M. V., Barros, J. C., Koller, W. W., & Andreotti, R. (2016). Avaliação da resistência de *Rhipicephalus* (*Boophilus*) *microplus* (Acari: Ixodidae) a diferentes formulações acaricidas utilizando amostras de propriedades brasileiras. *Revista Brasileira de Parasitologia Veterinaria, 25*(2), 163–171. https://doi.org/10.1590/S1984-29612016026

De Meneghi, D., Stachurski, F., & Adakal, H. (2016). Experiences in tick control by acaricide in the traditional cattle sector in Zambia and Burkina Faso: Possible environmental and public health implications. *Frontiers in Public Health, 4*(NOV). https://doi.org/10.3389/FPUBH.2016.00239

Dos Santos, D. S., Boito, J. P., Santos, R. C. V., Quatrin, P. M., Ourique, A. F., dos Reis, J. H., Gebert, R. R., Glombowsky, P., Klauck, V., Boligon, A. A., Baldissera, M. D., & Da Silva, A. S. (2017). Nanostructured cinnamon oil has the potential to control *Rhipicephalus microplus* ticks on cattle. *Experimental and Applied Acarology, 73*(1), 129–138. https://doi.org/10.1007/s10493-017-0171-5

Duan, Y., Dhar, A., Patel, C., Khimani, M., Neogi, S., Sharma, P., Siva Kumar, N., & Vekariya, R. L. (2020). A brief review on solid lipid nanoparticles: Part and parcel of contemporary drug delivery systems. *RSC Advances, 10*(45), 26777–26791. https://doi.org/10.1039/d0ra03491f

Elmowafy, M., Shalaby, K., Elkomy, M. H., Alsaidan, O. A., Gomaa, H. A. M., Abdelgawad, M. A., & Mostafa, E. M. (2023). Polymeric nanoparticles for delivery of natural bioactive agents: Recent advances and challenges. *Polymers, 15*(5), 1123. https://doi.org/10.3390/POLYM15051123

Fantatto, R. R., Gomes, A. R., Constantini, J. V. C., Rodero, C. F., Chorilli, M., de Chagas, A. C., Melero, A., & Pietro, R. C. L. R. (2024). Development and evaluation of the Acaricidal activity of Xantan gum-based hydrogel and polymeric nanoparticles containing *Achyrocline satureioides* extract. *Gels, 10*(10), 658. https://doi.org/10.3390/GELS10100658

Fantatto, R. R., Constantini, J. V. C., Politi, F. A. S., Sorrechia, R., Medeiros, C. C. B., Luiz, M. T., Bechara, G. H., de Souza Chagas, A. C., Chorilli, M., & Pietro, R. C. L. R. (2025). Current tick control strategies and prospects for using nanotechnology as an efficient alternative – A review. *Veterinary Sciences, 12*(2), 163. https://doi.org/10.3390/VETSCI12020163

Farina, P., Giunti, G., Campolo, O., Maggi, F., Ricciardi, R., Lucchi, A., Canale, A., Pavela, R., Guedes, R. N. C., Desneux, N., & Benelli, G. (2024). Nano- and microformulated botanicals for managing ticks and mites of medical and veterinary importance: Past, present, and future. *Industrial Crops and Products, 222*, 119809. https://doi.org/10.1016/J.INDCROP.2024.119809

Farjadian, F., Ghasemi, A., Gohari, O., Roointan, A., Karimi, M., & Hamblin, M. R. (2019). Nanopharmaceuticals and nanomedicines currently on the market: Challenges and opportunities. *Nanomedicine, 14*(1), 93–126. https://doi.org/10.2217/nnm-2018-0120

Figueiredo, A., Anholeto, L. A., Cola, D. F., Fantatto, R. R., Santos, I. B., Gainza, Y. A., Sousa, G. A., Pickett, L. J., Fraceto, L. F., & de Chagas, A. C. S. (2022). Nanoformulations with synthetic and plant-derived compounds for cattle tick control. *Veterinary Parasitology, 309*. https://doi.org/10.1016/j.vetpar.2022.109756

Figueiredo, A., Anholeto, L. A., Cola, D. F., Fantatto, R. R., Gainza, Y. A., dos Santos, I. B., Viçozzi, G. P., Ávila, D. S., Fraceto, L. F., & de Chagas, A. C. S. (2023). Acaricides containing zein nanoparticles: A tool for a lower impact control of the cattle tick *Rhipicephalus microplus*. *Veterinary Parasitology, 318*. https://doi.org/10.1016/j.vetpar.2023.109918

Gadelhaq, S. M., Ibrahium, S. M., Abdel-Baki, A. A. S., Arafa, W. M., Al-Quraishy, S., Hassan, A. O., Abdelgelil, N. H., Ahmed, M., & Aboelhadid, S. M. (2024). Efficacy and safety of geranium-oregano-thymol formulations to control of dog tick *Rhipicephalus sanguineus* sensu lato under laboratory and field conditions. *Veterinary Parasitology, 327*, 110112. https://doi.org/10.1016/J.VETPAR.2023.110112

Galli, G. M., Volpato, V., Santos, R. C. V., Gebert, R. R., Quatrin, P., Ourique, A. F., Klein, B., Wagner, R., Tonin, A. A., Baldissera, M. D., Stefani, L. M., & Da Silva, A. S. (2018). Effects of essential oil of *Eucalyptus globulus* loaded in nanoemulsions and in nanocapsules on reproduction of cattle tick (*Rhipicephalus microplus*). *Archivos de Zootecnia, 67*(260), 494–498. https://doi.org/10.21071/az.v0i0.3879

Gamal, A., Aboelhadid, S. M., Abo El-Ela, F. I., Abdel-Baki, A. A. S., Ibrahium, S. M., et al. (2023). Synthesis of Carvacrol-loaded Invasomes nanoparticles improved Acaricide efficacy, cuticle invasion and inhibition of Acetylcholinestrase against hard ticks. *Microorganisms, 11*(3), 733. https://doi.org/10.3390/MICROORGANISMS11030733/S1

García-Ponce, R., Villarreal-Villarreal, J. P., Suárez, A. E. F., Vidal, G. H., García-Hernández, D. G., Hernández-Escareño, J. J., & Heya, M. S. (2025). Nanocarriers of natural and synthetic ixodicides, new alternatives against *Rhipicephalus microplus* (Acari: Ixodidae): A review. *Veterinary Parasitology, 337*, 110506. https://doi.org/10.1016/J.VETPAR.2025.110506

Grisi, L., Leite, R. C., et al. (2014). Reavaliação do potencial impacto econômico de parasitos de bovinos no Brasil. *Revista Brasileira de Parasitologia Veterinaria, 23*(2), 150–156. https://doi.org/10.1590/S1984-29612014042

Gupta, P. K., Bhandari, N., Shah, H., Khanchandani, V., Keerthana, R., Nagarajan, V., & Hiremath, L. (2019). An update on nanoemulsions using nanosized liquid in liquid colloidal systems. In *Nanoemulsions, properties, fabrications and applications* (Vol. 1, pp. 1–14).

Heya, M. S., Cordero-Díaz, A., Galindo-Rodríguez, S. A., Verde-Star, M. J., Sánchez-García, E., Villarreal-Villarreal, J. P., & Guillén-Meléndez, G. A. (2023). Overcoming tumor and mucosal barriers through active-loaded nanocarriers: Nanoparticles and exosomes. *Applied Nanoscience (Switzerland), 13*(6), 4485–4495. https://doi.org/10.1007/s13204-022-02724-y

Hillaireau, H., & Couvreur, P. (2009). Nanocarriers' entry into the cell: Relevance to drug delivery. *Cellular and Molecular Life Sciences, 66*(17), 2873–2896. https://doi.org/10.1007/s00018-009-0053-z

Ibrahium, S. M., Aboelhadid, S. M., Wahba, A. A., Farghali, A. A., Miller, R. J., Abdel-Baki, A. A. S., & Al-Quraishy, S. (2022a). Preparation of geranium oil formulations effective for control of phenotypic resistant cattle tick *Rhipicephalus annulatus*. *Scientific Reports, 12*(1), 1–13. https://doi.org/10.1038/S41598-022-14661-5

Ibrahium, S. M., Wahba, A. A., Farghali, A. A., Abdel-Baki, A.-A. S., Mohamed, S. A. A., Al-Quraishy, S., Hassan, A. O., & Aboelhadid, S. M. (2022b). Acaricidal activity of tea tree and lemon oil Nanoemulsions against *Rhipicephalus annulatus*. *Pathogens, 11*(12), 1506. https://doi.org/10.3390/PATHOGENS11121506

Ibrahium, S. M., Abdel-Baki, A. A. S., Al-Quraishy, S., Hassan, K. M., Hassan, A. O., Abdel-Rahim, M. M., Arafa, W. M., Aboelhadid, S. M., & Gadelhaq, S. M. (2024). Efficacy of d-limonene Nanoemulsion against *Rhipicephalus annulatus* and *Rhipicephalus sanguineus* ticks. *Acta Parasitologica, 69*(1), 267–274. https://doi.org/10.1007/S11686-023-00734-6/METRICS

Irache, J. M., Esparza, I., Gamazo, C., Agüeros, M., & Espuelas, S. (2011). Nanomedicine: Novel approaches in human and veterinary therapeutics. *Veterinary Parasitology, 180*(1–2), 47–71. https://doi.org/10.1016/j.vetpar.2011.05.028

Jain, P., Satapathy, T., & Pandey, R. K. (2020). *Rhipicephalus microplus*: A parasite threatening cattle health and consequences of herbal acaricides for upliftment of livelihood of cattle rearing communities in Chhattisgarh. In *Biocatalysis and agricultural biotechnology* (Vol. 26). Elsevier Ltd. https://doi.org/10.1016/j.bcab.2020.101611

Kapoor, A., & Preet, S. (2023). Evaluation of Acaricidal activity of *Cinnamomum camphora* (F. Lauraceae) essential oil Nanoemulsion against cattle tick *Rhipicephalus microplus*. *Advances in Zoology and Botany, 11*(2), 121–128. https://doi.org/10.13189/AZB.2023.110204

Kaufman, W. R. (2013). Integument and ecdysis. In *Biology of ticks* (Vol. 1, 2nd ed., pp. 99–121).

Kreuter, J. (2007). Nanoparticles – A historical perspective. *International Journal of Pharmaceutics, 331*(1), 1–10. https://doi.org/10.1016/j.ijpharm.2006.10.021

Lammari, N., Louaer, O., Meniai, A. H., & Elaissari, A. (2020). Encapsulation of essential oils via nanoprecipitation process: Overview, progress, challenges and prospects. *Pharmaceutics, 12*(5). https://doi.org/10.3390/pharmaceutics12050431

Lazcano Díaz, E., Padilla Camberos, E., Castillo Herrera, G. A., Estarrón Espinosa, M., Espinosa Andrews, H., Paniagua Buelnas, N. A., Gutiérrez Ortega, A., & Martínez Velázquez, M. (2019). Development of essential oil-based phyto-formulations to control the cattle tick *Rhipicephalus microplus* using a mixture design approach. *Experimental Parasitology, 201*, 26–33. https://doi.org/10.1016/j.exppara.2019.04.008

Molaba, G. G., Thethe, P. K., Shai, L. J., Thekisoe, O. M. M., & Mtshali, K. (2025). A systematic literature review of the efficacy of acaricidal plants, plant extracts, and tick assays used against dog-associated ticks. *Acta Tropica, 265*, 107608. https://doi.org/10.1016/J.ACTATROPICA.2025.107608

Mulla, M. S., & Su, T. (1999). Activity and biological effects of neem products against arthropods of medical and veterinary importance. *Journal of the American Mosquito Control Association, 15*(2), 133–152.

Muqaddas, H., Mehmood, N., Rasool, M., Yousaf, F., & Iqbal, F. (2025). Tick-borne diseases. *Acaricides Resistance in Ticks*, 143–168. https://doi.org/10.1007/978-981-97-7411-1_7

Nawaz, S., Shamaila, S., Oneeb, M., Hassan, J. U., Ahmad, S., Ramakarshina, S., Sundarrajan, S., & Ban, D. (2025). Biocontrol potential of nano-encapsulated essential oil crosslinked with chitosan nanoparticles against tick-borne diseases. *Colloids and Surfaces A: Physicochemical and Engineering Aspects, 720*, 137166. https://doi.org/10.1016/J.COLSURFA.2025.137166

Nogueira, J. A. P., Figueiredo, A., Duarte, J. L., de Almeida, F. B., Santos, M. G., Nascimento, L. M., Fernandes, C. P., Mourão, S. C., Toscano, J. H. B., Rocha, L. M., & Chagas, A. C. S. (2020). Repellency effect of *Pilocarpus spicatus* A. St.-Hil essential oil and nanoemulsion against *Rhipicephalus microplus* larvae. *Experimental Parasitology, 215*. https://doi.org/10.1016/j.exppara.2020.107919

Paradise, J. J. (2019). Regulating nanomedicine at the food and drung administration. *AMA Journal of Ethics, 21*(4), 347–355.

Pardeike, J., Hommoss, A., & Müller, R. H. (2009). Lipid nanoparticles (SLN, NLC) in cosmetic and pharmaceutical dermal products. *International Journal of Pharmaceutics, 366*(1–2), 170–184. https://doi.org/10.1016/j.ijpharm.2008.10.003

Pavela, R. (2014). Insecticidal properties of *Pimpinella anisum* essential oils against the *Culex quinquefasciatus* and the non-target organism *Daphnia magna*. *Journal of Asia-Pacific Entomology, 17*(3), 287–293. https://doi.org/10.1016/j.aspen.2014.02.001

Pavela, R., & Benelli, G. (2016a). Essential oils as ecofriendly biopesticides? Challenges and constraints. *Trends in Plant Science, 21*(12), 1000–1007. https://doi.org/10.1016/j.tplants.2016.10.005

Pavela, R., & Benelli, G. (2016b). Ethnobotanical knowledge on botanical repellents employed in the African region against mosquito vectors – A review. *Experimental Parasitology, 167*, 103–108. https://doi.org/10.1016/j.exppara.2016.05.010

Pazinato, R., Klauck, V., Volpato, A., Tonin, A. A., Santos, R. C., de Souza, M. E., Vaucher, R. A., Raffin, R., Gomes, P., Felippi, C. C., Stefani, L. M., & Da Silva, A. S. (2014). Influence of tea tree oil (*Melaleuca alternifolia*) on the cattle tick *Rhipicephalus microplus*. *Experimental and Applied Acarology, 63*(1), 77–83. https://doi.org/10.1007/s10493-013-9765-8

Pazinatto Boito, J., Santos, R. C., Vaucher, R. A., Raffin, R., Machado, G., Tonin, A. A., & Da Silva, A. S. (2016). Evaluation of tea tree oil for controlling *Rhipicephalus microplus* in dairy cows. *Veterinary Parasitology, 225*, 70–72. https://doi.org/10.1016/j.vetpar.2016.05.031

Peltonen, L. (2018). Practical guidelines for the characterization and quality control of pure drug nanoparticles and nano-cocrystals in the pharmaceutical industry. *Advanced Drug Delivery Reviews, 131*, 101–115. https://doi.org/10.1016/j.addr.2018.06.009

Remedio, R. N., Nunes, P. H., & Camargo-Mathias, M. I. (2015). The extensible integument of *Rhipicephalus sanguineus* female ticks in different feeding stages: A morphological approach. *Acta Zoologica, 96*(3), 319–327. https://doi.org/10.1111/azo.12079

Riehemann, K., Schneider, S. W., Luger, T. A., Godin, B., Ferrari, M., & Fuchs, H. (2009). Nanomedicine – Challenge and perspectives. *Angewandte Chemie – International Edition, 48*(5), 872–897. https://doi.org/10.1002/anie.200802585

Rodríguez-Vivas, R. I., Trees, A. J., Rosado-Aguilar, J. A., Villegas-Perez, S. L., & Hodgkinson, J. E. (2011). Evolution of acaricide resistance: Phenotypic and genotypic changes in field populations of *Rhipicephalus* (*Boophilus*) *microplus* in response to pyrethroid selection pressure. *International Journal for Parasitology, 41*(8), 895–903. https://doi.org/10.1016/j.ijpara.2011.03.012

Rodríguez-Vivas, R. I., Grisi, L., Pérez de León, A. A., et al. (2017). Potential economic impact assessment for cattle parasites in Mexico. Review Evaluación del impacto económico potencial de los parásitos. *Revista Mexicana de Ciencias Pecuarias, 8*(1), 61–74.

Rodríguez-Vivas, R. I., España, E. R., Blanco, I. L., Ojeda-Chi, M. M., Trinidad-Martinez, I., Islas, J. A. T., & Bhushan, C. (2021). Monitoring the resistance of *Rhipicephalus microplus* to amitraz, flumethrin, coumaphos, and ivermectin on cattle farms in Mexico. *Veterinary Parasitology: Regional Studies and Reports, 26*. https://doi.org/10.1016/j.vprsr.2021.100644

Sales, A. M. R., Pereira, G. R. S., Lima, L. C. N., Monteiro, C. M. O., Matos, B. N., Taveira, S. F., Cunha-Filho, M., Gelfuso, G. M., & Gratieri, T. (2024). Thymol-loaded polymeric Nanocapsules' repellent activity on nymphs of *Rhipicephalus sanguineus* Sensu Lato. *Coatings, 14*(10), 1295. https://doi.org/10.3390/COATINGS14101295

Salvia-Trujillo, L., Soliva-Fortuny, R., Rojas-Graü, M. A., McClements, D. J., & Martín-Belloso, O. (2017). Edible Nanoemulsions as carriers of active ingredients: A review. *Annual Review of Food Science and Technology, 8*, 439–466. https://doi.org/10.1146/annurev-food-030216-025908

Shahzad, K., & Manzoor, F. (2019). Nanoformulations and their mode of action in insects: A review of biological interactions. *Drug and Chemical Toxicology, 44*(1), 1–11. https://doi.org/10.1080/01480545.2018.1525393

Sun, Y., Chen, D., Pan, Y., Qu, W., Hao, H., Wang, X., Liu, Z., & Xie, S. (2019). Nanoparticles for antiparasitic drug delivery. *Drug Delivery, 26*(1), 1206–1221. https://doi.org/10.1080/10717544.2019.1692968

Tienda-Vázquez, M. A., Melchor-Martínez, E. M., Elizondo-Luévano, J. H., Parra-Saldívar, R., Lara-Ortiz, J. S., Luna-Sosa, B., & Scheckhuber, C. Q. (2023). Antidiabetic plants for the treatment of type 2 diabetes mellitus and associated bacterial infections. *PRO, 11*(5), 1299. https://doi.org/10.3390/PR11051299

Tinkle, S., Mcneil, S. E., Mühlebach, S., Bawa, R., Borchard, G., Barenholz, Y. C., Tamarkin, L., & Desai, N. (2014). Nanomedicines: Addressing the scientific and regulatory gap. *Annals of the New York Academy of Sciences, 1313*(1), 35–56. https://doi.org/10.1111/nyas.12403

Tortorici, S., Cimino, C., Ricupero, M., Musumeci, T., Biondi, A., Siscaro, G., Carbone, C., & Zappalà, L. (2022). Nanostructured lipid carriers of essential oils as potential tools for the sustainable control of insect pests. *Industrial Crops and Products, 181*, 114766. https://doi.org/10.1016/J.INDCROP.2022.114766

Valsoni, L. M., de Freitas, M. G., Echeverria, J. T., Borges, D. G. L., Tutija, J., & de Borges, F. A. (2020). Resistance to all chemical groups of acaricides in a single isolate of *Rhipicephalus microplus* in Mato Grosso do Sul, Brazil. *International Journal of Acarology, 46*(4), 276–280. https://doi.org/10.1080/01647954.2020.1765867

Valsoni, L. M., de Freitas, M. G., Borges, D. G. L., & de Borges, F. A. (2021). Status of *Rhipicephalus microplus* resistance to ivermectin, fipronil and fluazuron in Mato Grosso do Su, Brazill. *Revista Brasileira de Parasitologia Veterinaria, 30*(1), 1–7. https://doi.org/10.1590/s1984-296120201091

Ventola, C. L. (2017). Progress in Nanomedicine: Approved and investigational Nanodrugs. *Pharmacy and Therapeutics, 42*(12), 742.

Vilela, V. L. R., Feitosa, T. F., Bezerra, R. A., Klafke, G. M., & Riet-Correa, F. (2020). Multiple acaricide-resistant *Rhipicephalus microplus* in the semi-arid region of Paraíba state, Brazil. *Ticks and Tick-borne Diseases, 11*(4). https://doi.org/10.1016/j.ttbdis.2020.101413

Zainab, R., Hasnain, M., Ali, F., Abideen, Z., Siddiqui, Z. S., Jamil, F., Hussain, M., & Park, Y. K. (2024). Prospects and challenges of nanopesticides in advancing pest management for sustainable agricultural and environmental service. *Environmental Research, 261*, 119722. https://doi.org/10.1016/J.ENVRES.2024.119722

Zolnik, B., Potter, T. M., & Stern, S. T. (2011). Detecting reactive oxygen species in primary hepatocytes treated with nanoparticles. In *Methods in molecular biology* (Vol. 697, pp. 173–179). Humana Press Inc.. https://doi.org/10.1007/978-1-60327-198-1_18

Chapter 10
Optimizing the Therapeutic Action of Natural Compounds Using Nanoparticles

Manuel López-Cabanillas Lomelí, Blanca Edelia González Martínez, María Julia Verde-Star, María Concepción Arellano Salazar, and Michel Stéphane Heya

1 Introduction

Traditional medicine continues to play a vital role worldwide, with the World Health Organization (WHO) estimating that about 80% of people globally rely on it as their primary treatment for diseases affecting the respiratory, renal, dermatological, inflammatory, and gastrointestinal systems (Vega-Menchaca et al., 2013). Extensive in vitro and in vivo research has revealed that many plant species exhibit important pharmacological activities, including antimicrobial, antioxidant, analgesic, and anticancer effects (Al-Haj et al., 2019; Heya et al., 2022, 2024; Farahmand et al., 2016).

Despite these promising results, significant challenges are involved in translating these benefits into clinical practice. These include toxicity related to solvents, instability under physiological conditions, variations in pH, poor bioavailability at target sites, environmental influences, and limitations inherent to traditional dosage forms (Armendáriz-Barragán et al., 2016; López-Cabanillas Lomelí et al., 2025). Moreover, over 70% of new drugs have poor water solubility, which restricts their absorption and reduces therapeutic efficacy (Rahman et al., 2020; Hauser et al., 2021). To address these issues, there is a pressing need to develop innovative drug-delivery systems that improve the therapeutic use of natural compounds. Nanotechnology has emerged as a particularly promising approach, allowing the production of nanoparticles (NPs) sized between 50 and 500 nm, loaded with plant

M. L.-C. Lomelí · B. E. G. Martínez · M. C. A. Salazar · M. S. Heya (✉)
Facultad de Salud Pública y Nutrición, Universidad Autónoma de Nuevo León,
Nuevo León, Mexico
e-mail: michel.heyax@uanl.edu.mx

M. J. Verde-Star
Facultad de Ciencias Biológicas, Universidad Autónoma de Nuevo León,
Nuevo León, Mexico

M. S. Heya (ed.), *Natural Alternatives and Nanotechnology Applied to the One Health Approach*, https://doi.org/10.1007/978-3-032-03675-9_10

extracts to boost their biological effectiveness. Recent studies indicate that silver nanoparticles (AgNPs) synthesized using leaf extracts of *Justicia glauca* possess strong antimicrobial properties, especially against bacteria and fungi relevant to human health. Additionally, liposomes containing curcumin have been shown to restore homeostasis in dental-pulp stem cells (Sinjari et al., 2019; Wong et al., 2019). Likewise, celastrol encapsulated in poly(ε-caprolactone) NPs has exhibited potential anticancer activity targeting prostate cancer (Shi et al., 2020). Research by Pacheco-Ordaz reports that polymeric NPs carrying crude extracts of *Berberis vulgaris* and *Curcuma longa* demonstrate effective biological activity against *Entamoeba histolytica* (Pacheco-Ordaz, 2018). Furthermore, quercetin delivered via polymeric NPs shows enhanced antioxidant capacity compared with the free compound (Jurica et al., 2017). By fine-tuning characteristics such as the material composition (organic, inorganic, or hybrid), particle shape (spherical, rod-like, or cubic), size (nano, meso, or macro), and surface properties (charge, functional groups, or targeting ligands), the therapeutic performance of nanocarriers can be optimized. Hence, nanotechnology stands out as a key tool for advancing medical treatments (Din et al., 2017; Heya et al., 2022).

In summary, this chapter underscores the importance of integrating nanotechnological advances with the therapeutic potential of natural products to enhance their pharmacological effectiveness and clinical application.

2 Biomedical Applications of Green Nanoparticles

In recent years, NPs have emerged as a technology with significant potential for clinical applications. This has led to the development of nanomedicine, a medical field that utilizes the principles and tools of nanotechnology for the prevention and treatment of various diseases. This discipline covers a wide range of human conditions, including those related to dentistry, dermatology, the digestive system, and systemic illnesses.

2.1 Antimicrobial Effect

Microbial diseases remain a substantial challenge to global public health, impacting over 900 million individuals with dermatological conditions (Hay et al., 2014). The escalating resistance of pathogenic microorganisms to conventional antimicrobial agents has spurred renewed interest in exploiting bioactive natural compounds for their therapeutic and antimicrobial efficacy (Pérez-Narváez et al., 2019). Notably, cashew-nut-shell derivatives such as anacardic acid and cardol exhibit significant antimicrobial activity against bacterial and fungal pathogens implicated in skin infections. Incorporation of these bioactives into nanostructured delivery systems, including chitosan and alginate-based NPs, enhances their physicochemical stability and facilitates controlled-release kinetics, thereby extending their antimicrobial activity (Filho et al., 2019). In vitro investigations demonstrate that these nanoencapsulated formulations outperform their free counterparts, achieving effective inhibition and eradication of pathogenic strains responsible for dermatophytosis and candidiasis (Filho et al., 2019). Furthermore, nanotechnology has been successfully applied to parasitic diseases, exemplified by the encapsulation of *Artemisia absinthium* essential oil within nanocochleates. This approach significantly potentiated antileishmanial efficacy in murine models infected with *Leishmania amazonensis*, as evidenced by substantial reductions in lesion dimensions and parasitic burden following intralesional administration, outperforming treatments with unencapsulated essential oil (Pérez-Narváez et al., 2019).

In addition to bacterial and parasitic infections, nanotechnology-based delivery systems have shown considerable promise in addressing viral infections affecting other organs. For instance, respiratory viral infections, such as those caused by influenza and respiratory syncytial virus (Jannat et al., 2021; Huang et al., 2024), pose major health burdens globally. NPs formulated with natural antiviral compounds, including flavonoids and alkaloids, have demonstrated enhanced antiviral efficacy by improving bioavailability and targeted delivery to infected pulmonary tissues (Jannat et al., 2021; Huang et al., 2024). Recent studies indicate that liposomal encapsulation of quercetin significantly reduces viral replication and inflammatory responses in murine models of influenza, thereby mitigating lung-tissue damage and improving survival rates (Jannat et al., 2021; Huang et al., 2024). This controlled delivery system not only potentiates the antiviral activity but also minimizes systemic toxicity, positioning nanotechnology as a versatile platform in the fight against viral diseases beyond the skin.

These results highlight nanotechnology as a promising platform for enhancing the therapeutic potential of natural products, facilitating the development of more efficacious and safer interventions against cutaneous and parasitic infections. This strategy also aligns with the One Health framework for 2030 by promoting integrative solutions that concurrently advance human, animal, and environmental health through sustainable and innovative therapeutic modalities (European Centre for Disease Prevention and Control, 2024).

2.2 Anti-Inflammatory and Antioxidant Activity

Inflammatory diseases and oxidative stress are interrelated pathophysiological processes that contribute to the development and progression of numerous chronic pathologies. In this context, bioactive compounds derived from medicinal plants, such as polyphenols and curcuminoids, have demonstrated promising antioxidant and anti-inflammatory properties. However, their clinical application is limited by poor solubility and bioavailability. Nanotechnology, particularly the development of nanoemulsions and liposomal NPs, has revolutionized the efficient and controlled delivery of these compounds, enhancing their biological activity (Baccarin et al., 2015; Wang et al., 2018).

Nanoemulsions have emerged as promising nanocarrier systems owing to their ability to enhance the stability, bioavailability, and targeted delivery of bioactive plant compounds, particularly those rich in polyphenols. Baccarin et al. (2015) developed two distinct nanoemulsions incorporating the ethyl-acetate fraction of pomegranate peel extract, utilizing pomegranate-seed oil and medium-chain triglycerides (MCTs) as the lipid phases. Both formulations presented particle sizes close to 200 nm, which is within the optimal range for cellular uptake and systemic circulation stability. The nanoemulsions exhibited strong physical stability, as indicated by their negative zeta potentials and low polydispersity indexes, factors that are crucial for preventing particle aggregation and ensuring homogeneous dispersion. Functionally, these nanoemulsions demonstrated significant antioxidant activity by inhibiting hemolysis induced by hydrogen peroxide (H_2O_2), with IC_{50} values of 27.16 μg/mL and 76.05 μg/mL for the pomegranate seed oil and MCT formulations respectively. This antioxidant effect was further substantiated by their ability to protect human erythrocyte membrane proteins from oxidative damage mediated by the azo compound AAPH, with protection ranging between 59.5% and 88.2% depending on the protein band and emulsion type. Such protection against oxidative membrane damage is critical in preventing hemolytic anemia and other oxidative stress-related pathologies. Moreover, the nanoemulsions displayed notable photoprotective properties in HaCat keratinocyte cells exposed to UVB radiation, evidencing a dose-dependent protective effect and an appreciable sun protection factor. This finding positions these nanoformulations as potential candidates for dermatological applications such as sunscreens and skin-repair products. Importantly, no phototoxic effects were observed in neutral red uptake assays conducted on Swiss albino mouse 3 T3 fibroblasts at concentrations around 25 μg/mL, suggesting a favorable safety profile for topical use (Baccarin et al., 2015).

Separately, curcumin, the principal bioactive constituent of *Curcuma longa*, has garnered substantial interest for its anti-inflammatory and antioxidant properties, although its precise molecular mechanisms remain to be fully elucidated. Sinjari et al. (2019) investigated the effects of curcumin encapsulated in liposomes (CurLIP) on dental-pulp stem cells, particularly in the context of inflammation induced by 2-hydroxyethyl methacrylate. At concentrations of 3 and 5 mmol/L, CurLIP treatment was shown to restore cellular homeostasis by reducing aberrant cell proliferation, inducing morphological differentiation, and modulating key inflammatory cytokines such as interleukin-6 (IL-6), interleukin-8 (IL-8), monocyte chemoattractant protein-1 (MCP1), and interferon gamma (IFNγ). This nanocarrier-based approach enables the targeted delivery and controlled release of curcumin, overcoming its intrinsic low bioavailability and instability under physiological conditions. The ability of CurLIP to influence multiple biochemical signaling pathways suggests a promising strategy for endodontic therapies aimed at mitigating inflammation and promoting tissue regeneration. Consequently, this innovation has significant potential to enhance dental care outcomes and contribute positively to public health (Sinjari et al., 2019).

2.3 Antitumoral Activity

Tumors represent a critical area of investigation for understanding the local absorption, distribution, and efficacy of therapeutic agents, particularly when delivered in NP form versus their free-drug counterparts. The enhanced permeability and retention (EPR) effect, which arises because of the abnormal vasculature and impaired lymphatic drainage in tumors, allows NPs to preferentially accumulate within malignant tissues. Quantitative studies have demonstrated that approximately 10–15% of administered NPs localize in tumor sites, a stark contrast to the 0.1% tumor uptake observed for conventional small-molecule drugs (Scheetz et al., 2019; Heya et al., 2022). This marked increase is largely attributed to passive transport mechanisms governed by particle size and the unique pathophysiological properties of tumor vasculature.

Nonetheless, NP delivery faces significant challenges related to the complex tumor microenvironment (TME). The TME is characterized by heterogeneous and often abnormal vascularization, elevated interstitial fluid pressure, and a densely crosslinked extracellular matrix (ECM), all of which present formidable barriers to deep tissue penetration and uniform drug distribution (Dai et al., 2018; Heya et al., 2022). These physiological impediments reduce therapeutic efficacy by limiting NP diffusion beyond perivascular regions, thereby necessitating tailored NP design to improve transport dynamics within the tumor mass. Surface charge is a critical physicochemical property influencing NP interactions within the TME. Negatively charged NPs exhibit enhanced penetration through the tumor ECM, likely due to electrostatic repulsion from the predominantly anionic components of the matrix, which reduces nonspecific binding and promotes diffusivity (Prausnitz and Langer, 2008; Heya et al., 2022). Conversely, positively charged particles often display increased cellular uptake but suffer from rapid clearance and toxicity concerns.

Particle-size optimization is another key factor. NPs of between 5 and 150 nm have been reported to possess prolonged systemic circulation, reduced renal clearance, and enhanced accumulation in tumor tissues, capitalizing on the EPR effect (Toy et al., 2014; Heya et al., 2022). Particles smaller than 5 nm tend to be rapidly cleared renally, whereas those exceeding 150 nm may be sequestered by the mononuclear phagocyte system, limiting tumor access. Moreover, NP shape profoundly affects biodistribution and cellular interactions. Flattened or disk-shaped NPs have demonstrated longer circulation times than spherical particles owing to altered hydrodynamic properties and reduced recognition by immune cells. In contrast, elongated (rod-shaped) NPs favor multivalent interactions with tumor cells and endothelial linings, potentially enhancing retention and internalization within the tumor milieu (Toy et al., 2014; Heya et al., 2022).

To further increase tumor specificity, recent advancements have focused on functionalizing NP surfaces by targeting ligands such as peptides, antibodies, or aptamers. Such surface modifications enable active targeting by binding to overexpressed receptors or antigens on tumor cells, facilitating receptor-mediated endocytosis and intracellular delivery (Nizzero et al., 2018). For example, mesoporous silica NPs conjugated with polyethylene glycol and hyaluronic acid have been employed for controlled and sustained release of chemotherapeutics such as doxorubicin. This design improves drug bioavailability, minimizes systemic toxicity, and enhances therapeutic efficacy by exploiting the affinity of hyaluronic acid for CD44 receptors frequently overexpressed in cancer cells (Zhang et al., 2019). Similarly, elongated metallic NPs conjugated with antibodies have shown more efficient tumor uptake relative to spherical counterparts, potentially due to superior margination dynamics within tumor vasculature and enhanced receptor-mediated binding (Heya et al., 2022). These advances underscore the importance of rational NP engineering, where size, shape, charge, and surface chemistry are synergistically optimized to overcome the multifaceted barriers posed by the TME.

In summary, the interplay between the physicochemical properties of NPs and tumor biology dictates the ultimate fate and efficacy of nanomedicines. Continued research efforts are vital to develop next-generation NPs capable of traversing the TME efficiently, ensuring homogenous drug distribution and maximizing therapeutic outcomes.

2.4 Anticancer Activity: Colorectal Cancer

Phytochemicals derived from medicinal plants have demonstrated promising chemopreventive and therapeutic effects in colorectal cancer (CRC) (López-Cabanillas Lomelí et al., 2025). However, their clinical application remains limited owing to challenges such as poor water solubility, low bioavailability, and rapid metabolic degradation (Sun et al., 2023). Nanotechnology has emerged as an effective strategy to overcome these limitations by enabling controlled, site-specific delivery and enhancing the pharmacokinetic profile of these bioactive compounds.

Recent studies have confirmed that NP formulations significantly enhance the antitumor potential of plant-derived extracts. For instance, AgNPs synthesized with *Vitex negundo* extract exhibited potent cytotoxic effects against HCT15 CRC cells, with an IC_{50} value of 20 μg/mL, indicating strong inhibition of cellular proliferation (Kumari Singh et al., 2024). Similarly, nanoformulations of polyphenols such as resveratrol and quercetin have shown improved intestinal absorption and increased cytotoxicity compared with their free-form counterparts (You et al., 2016). In a notable study, Cristina Mesas reported that calcium phosphate NPs loaded with esculetin and euphorbetin exhibited selective cytotoxicity toward cancerous cells and achieved a 62% reduction in tumor volume in murine models, underscoring their therapeutic efficacy (Zambrano et al., 2018). Moreover, poly(lactic-co-glycolic acid) NPs loaded with blueberry extract improved intestinal permeability and downregulated vascular endothelial growth factor (VEGF), a key mediator of tumor angiogenesis (López-Cabanillas Lomelí et al., 2025; Wang et al., 2019).

In vivo studies have further validated the enhanced efficacy of nanoformulated compounds. Gou et al. demonstrated that curcumin-loaded micelles exhibited significantly greater antitumor activity than free curcumin in CRC models (Gou et al.,

2011). Additionally, Xiang et al. reported a six-fold increase in the half-life of curcumin when delivered via NP encapsulation, highlighting the extended circulation time and sustained therapeutic action achieved through nanoformulation (Xiang et al., 2014). Furthermore, Aboulthana et al. found that NPs synthesized using *Croton tiglium* extract modulated critical oncogenes and tumor-suppressor genes, including *TP53* and *APC*, both of which are essential in colorectal carcinogenesis (Aboulthana et al., 2020). On the clinical front, a phase II trial combining bevacizumab with the FOLFIRI regimen demonstrated favorable safety and efficacy profiles, suggesting that nanotechnology-based strategies might be compatible with existing chemotherapeutic protocols (Jeon et al., 2022). Taken together, these findings underscore the transformative potential of nanotechnology in CRC treatment. By enhancing the bioactivity, targeting, and therapeutic window of plant-derived agents, NPs present a promising platform for developing more effective and safer alternatives to conventional chemotherapy.

2.5 Nephroprotective Activity

Metallic NPs enriched with medicinal plant derivatives, including gold nanoparticles (AuNPs), AgNPs, and metal oxide nanoparticles (MONPs), are emerging as promising therapeutic agents for the treatment and protection of renal function in a wide range of pathological conditions (Chaudhary et al., 2023; Fernandes, 2023; Tripathi and Singh, 2022). These nanostructures possess unique physicochemical properties such as high surface-area-to-volume ratios, tunable size, shape, and surface charge that confer enhanced biological activity and allow them to interact efficiently with cellular and molecular targets in the kidney. Their multifunctionality, combined with antioxidant, anti-inflammatory, and cytoprotective properties, positions them as innovative tools in nephroprotection and regenerative nephrology.

2.5.1 Gold Nanoparticles

AuNPs have garnered significant attention owing to their excellent biocompatibility, ease of surface functionalization, and potent biological activities. The nephroprotective effects of AuNPs are largely dependent on their physicochemical characteristics, particularly size, shape, and surface charge, which influence their biodistribution, cellular uptake, and interaction with renal tissues (Fernandes, 2023). Mechanistically, AuNPs exert antioxidant effects by scavenging reactive oxygen species (ROS) and enhancing the activity of endogenous antioxidant enzymes such as superoxide dismutase, catalase, and glutathione peroxidase (Tripathi and Singh 2022). In addition, AuNPs modulate pro-inflammatory cytokines (e.g., TNF-α, IL-6) and signaling pathways (e.g., NF-κB), thereby reducing inflammation and cellular apoptosis (Arcos Rosero et al., 2024; Tripathi and Singh, 2022). These effects have been demonstrated in models of acute kidney injury, nephrotoxicity induced by chemotherapeutic agents, and metabolic disorders such as diabetic nephropathy (Peres et al., 1867). Moreover, AuNPs have shown promise in preserving glomerular structure, maintaining tubular integrity, and attenuating fibrosis (Al-Waili et al., 2017).

2.5.2 Silver Nanoparticles

AgNPs, known for their broad-spectrum antimicrobial properties, also exhibit notable renoprotective potential. In preclinical studies, AgNPs have demonstrated the ability to mitigate oxidative stress by reducing lipid peroxidation and ROS generation (Khuda et al., 2023). They also modulate the expression of inflammatory mediators and promote renal tissue repair. Their therapeutic efficacy has been evaluated in diverse pathological contexts, including diabetic nephropathy, nephrolithiasis, and nephrotoxicity caused by environmental toxins or pharmaceuticals (Khuda et al., 2023; Qiu et al., 2023). AgNPs can restore the glomerular filtration rate, decrease creatinine and blood urea nitrogen levels, and promote the regeneration of damaged renal tissues. However, owing to their potential cytotoxicity at higher concentrations, careful optimization of dose and surface functionalization is necessary to ensure safety and efficacy (Wang and Zhang, 2025; Dighade et al., 2021).

2.5.3 Metal Oxide Nanoparticles

MONPs, such as zinc oxide (ZnO), iron oxide (Fe_2O_3 or Fe_3O_4), and cerium oxide (CeO_2) NPs, offer a wide array of therapeutic benefits for kidney diseases owing to their intrinsic redox activity and regenerative potential (Fernandes, 2023). ZnO NPs have been shown to reduce oxidative damage, restore zinc homeostasis, and enhance renal repair mechanisms. Iron-oxide NPs, particularly superparamagnetic variants, not only serve therapeutic purposes but also offer diagnostic value as contrast agents in magnetic resonance imaging (Hassan et al., 2021; Alomari et al., 2021). CeO_2 NPs exhibit regenerative redox cycling between Ce^{3+} and Ce^{4+} states, enabling continuous scavenging of ROS and prevention of lipid peroxidation (Meng et al., 2025; Khan et al., 2023); these NPs have demonstrated efficacy in reducing tubular damage, mitigating inflammatory responses, and improving renal biomarkers.

In summary, metallic NPs represent a novel and multifunctional therapeutic platform for addressing a wide range of renal pathologies, including acute kidney injury, diabetic nephropathy, nephrotoxicity, fibrosis, and chronic kidney disease. Their ability to modulate oxidative stress, inflammation, apoptosis, and fibrosis at the molecular level positions them as strong candidates for next-generation renal therapeutics. However, further clinical investigations are required to validate their safety, pharmacokinetics, and long-term efficacy. Integrating metallic NPs into renal medicine could mark a significant advancement in personalized nephrology and open new avenues for targeted, efficient, and safer treatment strategies.

2.6 Hepatoprotective Activity of Green Nanoparticles

The liver plays a central role in numerous metabolic functions and is highly susceptible to damage caused by viruses, alcohol consumption, poor dietary habits, and metabolic disorders. These factors can lead to severe pathologies such as cirrhosis and cancer (Asrani et al., 2019). Treating liver diseases remains a major challenge owing to their complex pathophysiology and progressive nature (Böttger et al., 2020). NPs loaded with plant-derived compounds have shown significant hepatoprotective potential by enhancing the delivery and efficacy of natural molecules with antioxidant and anti-inflammatory properties, such as curcumin (Fallowfield et al., 2021; Heya et al., 2023a, b). These nanosystems improve the stability and bioavailability of active compounds and allow for their controlled release specifically in the liver, thereby overcoming biological barriers and minimizing adverse effects (Fallowfield et al., 2021; Fu et al., 2021). As such, nanotechnology emerges as an innovative strategy for developing more effective and safer therapies for liver diseases, supporting both liver-tissue protection and regeneration (Na-Hyun Kim et al., 2017).

2.6.1 Fibrosis

Hepatic fibrosis is characterized by excessive collagen accumulation and the activation of hepatic stellate cells (HSCs), processes that drive inflammation and progression toward cirrhosis or hepatocellular carcinoma. The activation of HSCs, mediated by the overexpression of $\alpha v\beta 3$ integrins, represents a critical therapeutic target. Several studies have investigated the use of NPs to inhibit HSC activation and treat fibrosis. For instance, liposomal NPs functionalized with the cRGDyK peptide have shown high affinity for activated HSCs and efficient cellular internalization (Ciceu et al., 2024). Furthermore, silymarin encapsulated in chitosan NPs has been reported to downregulate key fibrogenic mediators such as TGFβR1 and COL3A1 while upregulating protective microRNAs in the liver. Similarly, polymeric NPs carrying curcumin have demonstrated the ability to inhibit the expression of fibrosis-related proteins and induce apoptosis in activated HSCs in experimental models. These approaches underscore the potential of NPs to improve the targeted delivery and therapeutic efficacy of antifibrotic agents (Ciceu et al., 2024).

2.6.2 Carcinoma

NPs also offer a novel therapeutic approach for advanced liver diseases such as cirrhosis and hepatocellular carcinoma, where physiological barriers limit drug access to critical cells such as hepatocytes and HSCs. Owing to their customizable physicochemical properties, these nanosystems enable controlled and selective drug release, facilitating access to more therapeutically accessible cells, such as Kupffer cells. Additionally, NPs can deliver hepatoprotective, antioxidant, and anti-inflammatory agents that not only reduce cellular damage but also promote the apoptosis of pathological cells and slow fibrosis progression. Their ability to overcome barriers such as the space of Disse and enhance liver regeneration positions them as fundamental tools in the development of precision therapies for chronic liver diseases, paving the way for more effective and targeted treatment strategies (Heya et al., 2023a, b).

3 Perspectives and Conclusion

The application of nanotechnology for the delivery of bioactive molecules in the treatment of human diseases presents a promising alternative to enhance the therapeutic efficacy of active compounds, increase their bioavailability, and enable controlled release at targeted sites. Nevertheless, considering nanotechnology solely from this standpoint would underestimate its broader potential in addressing chronic and complex pathologies. Recent advances in nanomedicine have facilitated the rational design of NPs with optimized physicochemical properties, enabling their integration into precision and personalized medicine paradigms. The development of NPs for therapeutic use has inaugurated a new frontier in biomedical science, positioning these nanocarriers as highly efficient vectors for the targeted delivery of therapeutic agents. Their ability to traverse biological barriers, along with their biocompatibility, low cytotoxicity, and stability under diverse physiological conditions, renders them pivotal tools, not only in personalized human medicine but also within a comprehensive health approach that encompasses animal and environmental well-being.

Aligned with the One Health 2030 Agenda, which advocates for a holistic vision of global health, NPs are emerging as a versatile platform with transdisciplinary applications beyond the human clinical context. Their intelligent design allows for adaptation to diverse biological systems, thereby facilitating their deployment in veterinary medicine, zoonotic disease control, and mitigation of the environmental impacts associated with pharmaceutical interventions. This integrative capacity addresses the growing demand for solutions that recognize the interconnectivity among human, animal, and ecosystem health.

Although biological heterogeneity across human and animal populations poses a significant challenge to the development of universal therapies, the rational engineering of NPs allows for customization to specific physiological and pathological contexts, thereby enhancing therapeutic efficacy at both individual and population levels. Moreover, their potential to reduce toxic residues and curb antimicrobial resistance positions them as strategic assets in the development of sustainable therapeutic strategies aligned with the One Health framework.

In summary, NPs represent not only a significant biomedical innovation but also a critical tool in advancing global commitments toward integrated and sustainable health. Their continued development and responsible application may be instrumental in achieving public health and ecological targets established for the year 2030.

References

Aboulthana, W. M., Ibrahim, N. E. S., Osman, N. M., Seif, M. M., Hassan, A. K., Youssef, A. M., El-Feky, A. M., & Madboli, A. A. (2020). Evaluation of the biological efficiency of silver nanoparticles biosynthesized using Croton tiglium L. seeds extract against azoxymethane induced colon cancer in rats. *Asian Pacific Journal of Cancer Prevention: APJCP, 21*(5), 1369.

Al-Haj, N., Reem, A., Al-Shamahy, H., Al-Moyed, K., Bahaj, S. S., & Jaber, A. (2019). Antimicrobial activity of five Yemeni medicinal plants against selected human pathogenic bacteria and fungi. *American Journal of Plant Sciences, 10*(September), 1699–1707. https://doi.org/10.4236/ajps.2019.110121

Alomari, G., Al-Trad, B., Hamdan, S., Aljabali, A. A. A., Al Zoubi, M. S., Al-Batanyeh, K., Qar, J., et al. (2021). Alleviation of diabetic nephropathy by zinc oxide nanoparticles in streptozotocin-induced type 1 diabetes in rats. *IET Nanobiotechnology, 15*(5), 473–483. https://doi.org/10.1049/nbt2.12026

Al-Waili, N., Al-Waili, H., Al-Waili, T., & Salom, K. (2017). Natural antioxidants in the treatment and prevention of diabetic nephropathy; a potential approach that warrants clinical trials. *Redox Report, 22*(3), 99–118. https://doi.org/10.1080/13510002.2017.1297885

Armendáriz-Barragán, B., Zafar, N., Badri, W., Galindo-Rodríguez, S., Kabbaj, D., Fessi, H., & Elaissari, A. (2016). Expert opinion on drug delivery plant extracts: From encapsulation to application. *Expert Opinion on Drug Delivery, 13*(8), 1165–1175. https://doi.org/10.1080/17425247.2016.1182487

Asrani, S. K., Devarbhavi, H., Eaton, J., & Kamath, P. S. (2019). Burden of liver diseases in the world. *Journal of Hepatology, 70*(1), 151–171. https://doi.org/10.1016/j.jhep.2018.09.014

Baccarin, T., Lemos-Senna, E., & Pilar, M. (2015). Toxicology in vitro protection against oxidative damage in human erythrocytes and preliminary Photosafety assessment of Punica Granatum seed oil Nanoemulsions entrapping polyphenol-rich ethyl acetate fraction. *TIV, 30*(1), 421–428. https://doi.org/10.1016/j.tiv.2015.09.020

Böttger, R., Pauli, G., Chao, P. H., Fayez, N. A. L., Hohenwarter, L., & Li, S. D. (2020). Lipid-based nanoparticle technologies for liver targeting. In *Advanced drug delivery reviews*. Elsevier B.V. https://doi.org/10.1016/j.addr.2020.06.017

Chaudhary, K. R., Banik, P., & Singh, K. (2023). Recent trends in the delivery of plant-derived phytochemicals against various cancers using Nanotechnological approach: A comprehensive review. *Journal of Drug Delivery Science and Technology, 87*, 104859. https://doi.org/10.1016/j.jddst.2023.104859

Ciceu, A., Fenyvesi, F., Hermenean, A., Ardelean, S., Dumitra, S., & Puticiu, M. (2024). Advancements in plant-based therapeutics for hepatic fibrosis: Molecular mechanisms and nanoparticulate drug delivery systems. *International Journal of Molecular Sciences, 25*(17). https://doi.org/10.3390/ijms25179346

Dai, Q., Wilhelm, S., Ding, D., Syed, A. M., Sindhwani, S., Zhang, Y., Chen, Y. Y., Macmillan, P., & Chan, W. C. W. (2018). Quantifying the ligand-coated nanoparticle delivery to cancer cells in solid tumors. *ACS Nano, 12*(8), 8423–8435. https://doi.org/10.1021/acsnano.8b03900

Dighade, R., Ingole, R., Ingle, P., Gade, A., Hajare, S., & Ingawale, M. (2021). Nephroprotective effect of Bryophyllum Pinnatum-mediated silver nanoparticles in ethylene glycol-induced Urolithiasis in rat. *IET Nanobiotechnology, 15*(3), 266–276. https://doi.org/10.1049/nbt2.12011

Din, F. U., Aman, W., Ullah, I., Qureshi, O. S., Mustapha, O., Shafique, S., & Zeb, A. (2017). Effective use of nanocarriers as drug delivery systems for the treatment of selected tumors. *International Journal of Nanomedicine, 12*, 7291–7309. https://doi.org/10.2147/IJN.S146315

European Centre for Disease Prevention and Control (ECDC). (2024). ECDC One Health Framework, no. May, 1–17. 10.2900/634973.

Fallowfield, J. A., Jimenez-Ramos, M., & Robertson, A. (2021). Emerging synthetic drugs for the treatment of liver cirrhosis. In *Expert opinion on emerging drugs*. Taylor and Francis Ltd. https://doi.org/10.1080/14728214.2021.1918099

Farahmand, S., Rasooli, A., & Saffarpour, M. (2016). Antifungal activities of methanolic extract of plants. *Electronic Journal of Biology, 1*, 42–44.

Fernandes, D. A. (2023). Review on metal-based Theranostic nanoparticles for cancer therapy and imaging. *Technology in Cancer Research and Treatment, 22*. https://doi.org/10.1177/15330338231191493

Filho, J. C., Paiva, S. M., Morais, D., Sobrinho, A. C. N., Cavalcante, G. S., Da Silva, N. A., & Abreu, F. O. M. D. S. (2019). Design of chitosan-alginate core-shell nanoparticules loaded with anacardic acid and cardol for drug delivery. *Polimeros, 29*(4), 1–10. https://doi.org/10.1590/0104-1428.08118

Fu, Y. S., Chen, T. H., Weng, L., Huang, L., Lai, D., & Weng, C. F. (2021). Pharmacological properties and underlying mechanisms of curcumin and prospects in medicinal potential. *Biomedicine and Pharmacotherapy, 141*(June), 111888. https://doi.org/10.1016/j.biopha.2021.111888

Gou, M., Men, K., Shi, H., Xiang, M., Zhang, J., Song, J., Long, J., et al. (2011). Curcumin-loaded biodegradable polymeric micelles for colon cancer therapy in vitro and in vivo. *Nanoscale, 3*(4), 1558–1567. https://doi.org/10.1039/c0nr00758g

Hassan, S. A., Mujahid, H., Ali, M. M., Irshad, S., Naseer, R., Saeed, S., Firyal, S., & Arooj, F. (2021). Synthesis, characterization and protective effect of green tea-mediated zinc oxide nanoparticles against Ochratoxin a induced hepatotoxicity and nephrotoxicity in albino rats. *Applied Nanoscience (Switzerland), 11*(8), 2281–2289. https://doi.org/10.1007/s13204-021-02006-z

Hauser, P. V., Chang, H. M., Yanagawa, N., & Hamon, M. (2021). Nanotechnology, nanomedicine, and the kidney. *Applied Sciences, 11*(16), 7187.

Hay, R. J., Johns, N. E., Williams, H. C., Bolliger, I. W., Dellavalle, R. P., Margolis, D. J., Marks, R., et al. (2014). The global burden of skin disease in 2010: An analysis of the prevalence and impact of skin conditions. *Journal of Investigative Dermatology, 134*(6), 1527–1534. https://doi.org/10.1038/jid.2013.446

Heya, M. S., García-Coronado, P. L., Cordero-Díaz, A., Torres-Hernández, A. D., & Pérez-Narváez, O. A. (2022). Biomedical application of nanoformulated plant. *International Journal of Pharmaceutical Sciences and Research, 13*(1), 1–12. https://doi.org/10.13040/IJPSR.0975-8232.13(1).1-12

Heya, M. S., Cordero-Díaz, A., Galindo-Rodríguez, S. A., Verde-Star, M. J., Sánchez-García, E., Villarreal-Villarreal, J. P., & Guillén-Meléndez, G. A. (2023a). Overcoming tumor and mucosal barriers through active-loaded nanocarriers: Nanoparticles and exosomes. *Applied Nanoscience, 13*(6), 4485–4495.

Heya, M. S., García-Ponce, R., Medina-Soto, B., Verde-Star, M., Soto-Domínguez, A., García-Hernández, D., Saucedo-Cardenas, O., Hernández-Salazar, M., & Guillen, G. (2023b). Green alternatives in treatment of liver diseases: The challenges of traditional medicine and green Nanomedicine. *Chemistry & Biodiversity, 20*(August), e202300463. https://doi.org/10.1002/cbdv.202300463

Heya, M. S., Verde-Star, M. J., Rivas-Morales, C., García-Hernández, D. G., Tijerina-Sáenz, A., López-Cabanillas-lomelí, M., Álvarez-Román, R., & Galindo-Rodríguez, S. A. (2024). In vitro antifungal activity of polymeric nanoparticles loaded with euphorbia Tirucalli extract. *Brazilian Journal of Biology, 84*. https://doi.org/10.1590/1519-6984.275974

Huang, M., Liu, X., Ren, Y., Huang, Q., Shi, Y., Yuan, P., & Chen, M. (2024). Quercetin: A flavonoid with potential for treating acute lung injury. *Drug Design, Development and Therapy, 18*(November), 5709–5728. https://doi.org/10.2147/DDDT.S499037

Jannat, K., Paul, A. K., Bondhon, T. A., Hasan, A., Nawaz, M., Jahan, R., Mahboob, T., et al. (2021). Nanotechnology applications of flavonoids for viral diseases. *Pharmaceutics, 13*(11), 1–37. https://doi.org/10.3390/pharmaceutics13111895

Jeon, Y., Sym, S. J., Yoo, B. K., & Baek, J. H. (2022). Long-term survival, tolerability, and safety of first-line bevacizumab and FOLFIRI in combination with ginsenoside-modified nanostructured lipid carrier containing curcumin in patients with unresectable metastatic colorectal cancer. *Integrative Cancer Therapies, 21.*

Jurica, K., Gobin, I., Kremer, D., Čepo, D. V., Grubešić, R. J., Karačonji, I. B., & Kosalec, I. (2017). Arbutin and its metabolite hydroquinone as the Main factors in the antimicrobial effect of strawberry tree (arbutus Unedo L.) leaves. *Journal of Herbal Medicine, 8*, 17–23. https://doi.org/10.1016/j.hermed.2017.03.006

Khan, M., Sohail, N. I., Raja, M. J., et al. (2023). Antioxidant and hypoglycemic potential of phytogenic cerium oxide nanoparticles. *Scientific Reports, 13*(1), 1–13. https://doi.org/10.1038/s41598-023-31498-8

Khuda, F., Gul, M., Khalil, A. A. K., Ali, S., Ullah, N., Khan, M. S., Nazir, S., et al. (2023). Biosynthesized silver nanoparticles using Alnus Nitida leaf extract as a potential antioxidant and anticancer agent. *ACS Omega, 8*(33), 30221–30230. https://doi.org/10.1021/acsomega.3c02928

Kim, N. H., Heo, J. D., Rho, J. R., & Yang, M. H. (2017). The standardized extract of Limonium Tetragonum alleviates chronic alcoholic liver injury in C57BL/6J mice. *Pharmacognosy Magazine, 13*(62), 179–188. https://doi.org/10.4103/pm.pm

López-Cabanillas Lomelí, M., Tijerina-Sáenz, A., & García-Hernández, D. G. (2025). Colon cancer: Overview on improved therapeutic potential of plant-based compounds using nanotechnology. *Scientia Pharmaceutica, 93*(1). https://doi.org/10.3390/scipharm93010001

Meng, Z., Tang, M., Shiyin, X., Zhou, X., Zhang, Z., Yang, L., Nüssler, A. K., Liu, L., & Yang, W. (2025). Protective effects of bone marrow mesenchymal stem cell-derived exosomes loaded cerium dioxide nanoparticle against Deoxynivalenol-induced liver damage. *Journal of Nanobiotechnology, 23*(1). https://doi.org/10.1186/s12951-025-03316-y

Nizzero, S., Ziemys, A., & Ferrari, M. (2018). Transport barriers and Oncophysics in cancer treatment. *Trends in Cancer, 4*(4), 277–280. https://doi.org/10.1016/j.trecan.2018.02.008

Pacheco-Ordaz, A. (2018). *Actividad Antiparasitaria in Vitro de Extractos de Curcuma Longa Y Berberis Vulgaris Incorporados En Nanoparticulas Poliméricas Contra Protozoarios de Interés Clínico*. Universidad Autónoma de Nuevo León.

Peres, R. A. S., Silva-Aguiar, R. P., Teixeira, D. E., Peruchetti, D. B., Alves, S. A. S., et al. (1867). Gold nanoparticles reduce tubule-interstitial injury and proteinuria in a murine model of subclinical acute kidney injury. *Biochimica et Biophysica Acta – General Subjects, 2023*(4). https://doi.org/10.1016/j.bbagen.2023.130314

Pérez-Narváez, O. A., Leos-Rivas, C., Rivas-Morales, C., Villarreal-Treviño, L., Barrón-González, M. P., & Sáncehz-García, E. (2019). Actividad Antimicrobiana y Antioxidante de Extractos Etanólicos de Oja de Arbutus Xalapensis Kunt, Mimosa Malacophylla Gray y Teucrium Cubense Jacquin. *Revista Tendencias En Docencia e Investigación En Química, 5*, 739–747.

Prausnitz, M. R., & Langer, R. (2008). Transdermal Drug Delivery. *Nature Biotechnology, 26*(11), 1261. https://doi.org/10.1038/NBT.1504

Qiu, F., Hulin, L., Wang, X., Yang, Y., & Ding, M. (2023). Evaluation of the Nephroprotective properties of silver nanoparticles green-mediated by Arabic gum on the lipopolysaccharide-induced acute kidney injury. *Inorganic Chemistry Communications, 155*, 111043. https://doi.org/10.1016/j.inoche.2023.111043

Rahman, H. S., Othman, H. H., Hammadi, N. I., Yeap, S. K., Amin, K. M., Samad, N. A., & Alitheen, N. B. (2020). Novel drug delivery systems for loading of natural plant extracts and their biomedical applications. *International Journal of Nanomedicine, 15*, 2439–2483. https://doi.org/10.2147/IJN.S227805

Rosero, W. A. A., Barbezan, A. B., de Souza, C. D., & Rostelato, M. E. C. M. (2024). Review of advances in coating and functionalization of gold nanoparticles: From theory to biomedical application. *Pharmaceutics, 16*(2), 255.

Scheetz, L., Park, K. S., Li, Q., Lowenstein, P. R., Castro, M. G., Schwendeman, A., & Moon, J. J. (2019). Engineering patient-specific cancer immunotherapies. *Nature Biomedical Engineering, 3*(10), 768–782. https://doi.org/10.1038/s41551-019-0436-x

Shi, J., Li, J., Xu, Z., Liang, C., Luo, R., Zhang, C., Gao, F., Zhang, J., & Chaomei, F. (2020). Celastrol: A review of useful strategies overcoming its limitation in anticancer application. *Frontiers in Pharmacology, 11*(November). https://doi.org/10.3389/fphar.2020.558741

Singh, S. K., Solanki, R., & Patel, S. (2024). Nano Pharmaceutical delivery in combating colorectal cancer. *Medicine in Drug Discovery, 21*(October 2023), 100173. https://doi.org/10.1016/j.medidd.2023.100173

Sinjari, B., Pizzicannella, J., D'Aurora, M., Zappacosta, R., Gatta, V., Fontana, A., Trubiani, O., & Diomede, F. (2019). Curcumin/liposome nanotechnology as delivery platform for anti-inflammatory activities via NFkB/ERK/PERK pathway in human dental pulp treated with 2-HydroxyEthyl MethAcrylate (HEMA). *Frontiers in Physiology, 10*(June), 1–11. https://doi.org/10.3389/fphys.2019.00633

Sun, L., Liu, H., Ye, Y., Lei, Y., Islam, R., Tan, S., Tong, R., Miao, Y. B., & Cai, L. (2023). Smart nanoparticles for cancer therapy. *Signal Transduction and Targeted Therapy, 8*(1). https://doi.org/10.1038/s41392-023-01642-x

Toy, R., Peiris, P. M., Ghaghada, K. B., & Karathanasis, E. (2014). Shaping cancer nanomedicine: The effect of particle shape on the in vivo journey of nanoparticles. *Nanomedicine, 9*(1), 121–134. https://doi.org/10.2217/nnm.13.191

Tripathi, M., & Singh, B. (2022). Chapter 6 – Synthesis of green nanocomposite material for engineering application. In J. R. Koduru et al. (Eds.), *Sustainable nanotechnology for environmental remediation* (pp. 135–157). Micro and Nano Technologies. Elsevier. https://doi.org/10.1016/B978-0-12-824547-7.00012-6

Vega-Menchaca, M., del Carmen, J., Verde-Star, A. O.-C., Morales-Rubio, M. E., Núñez-González, M. A., Rivera-Guillén, M. A., Serrano-Gallardo, L. B., & Rivas-Morales, C. (2013). Actividad Antibacteriana y Citotóxica de Leucophyllum Frutescens (Berl) I.M. Johnst Del Norte de México Contra Staphylococcus Aureus de Aislados Clínicos. *Revista Mexicana de Ciencias Farmaceuticas, 44*(2), 24–30.

Wang, Z., & Zhang, C. (2025). Nanomaterials for targeted therapy of kidney diseases: Strategies and advances. *Materials Today Bio, 31*(November 2024), 101534. https://doi.org/10.1016/j.mtbio.2025.101534

Wang, T., Qiaobin, H., Lee, J. Y., & Luo, Y. (2018). Solid lipid-polymer hybrid nanoparticles by in situ conjugation for oral delivery of Astaxanthin. *Journal of Agricultural and Food Chemistry, 66*(36), 9473–9480. https://doi.org/10.1021/acs.jafc.8b02827

Wang, X., Yang, C., Zhang, Q., Wang, C., Zhou, X., Zhang, X., & Liu, S. (2019). In vitro anticancer effects of Esculetin against human leukemia cell lines involves apoptotic cell death, autophagy, G0/G1 cell cycle arrest and modulation of Raf/MEK/ERK Signalling pathway. *Journal of the Balkan Union of Oncology, 24*(4), 1686–1691.

Wong, K. E., Ngai, S. C., Chan, K. G., Lee, L. H., Goh, B. H., & Chuah, L. H. (2019). Curcumin nanoformulations for colorectal cancer: A review. *Frontiers in Pharmacology, 10*, 152.

Xiang, D., Shigdar, S., Yang, W., Duan, W., Li, Q., Lin, J., Liu, K., & Li, L. (2014). Epithelial cell adhesion molecule Aptamer functionalized PLGA-lecithin-curcumin-PEG nanoparticles for targeted drug delivery to human colorectal adenocarcinoma cells. *International Journal of Nanomedicine*, 1083. https://doi.org/10.2147/IJN.S59779

You, X., Kang, Y., Hollett, G., Chen, X., Zhao, W., Zhipeng, G., & Jun, W. (2016). Polymeric nanoparticles for colon cancer therapy: Overview and perspectives. *Journal of Materials Chemistry B. Royal Society of Chemistry*. https://doi.org/10.1039/c6tb01925k

Zambrano, L. M. G., Brandao, D. A., Rocha, F. R. G., Marsiglio, R. P., Longo, I. B., Primo, F. L., Tedesco, A. C., Guimaraes-Stabili, M. R., & Rossa, C. (2018). Local administration of curcumin-loaded nanoparticles effectively inhibits inflammation and bone resorption associated with experimental periodontal disease. *Scientific Reports, 8*(1), 1–11. https://doi.org/10.1038/s41598-018-24866-2

Zhang, R., Qin, X., Kong, F., Chen, P., & Pan, G. (2019). Improving cellular uptake of therapeutic entities through interaction with components of cell membrane. *Drug Delivery, 26*(1), 328. https://doi.org/10.1080/10717544.2019.1582730

Index

M. S. Heya (ed.), *Natural Alternatives and Nanotechnology Applied to the One Health Approach*, https://doi.org/10.1007/978-3-032-03675-9

MIX
Papier aus verantwortungsvollen Quellen
Paper from responsible sources
FSC® C105338

If you have any concerns about our products,
you can contact us on
ProductSafety@springernature.com

In case Publisher is established outside the EU,
the EU authorized representative is:
Springer Nature Customer Service Center GmbH
Europaplatz 3, 69115 Heidelberg, Germany

Printed by Libri Plureos GmbH
in Hamburg, Germany